Advances in Asian Human-Environmental Research

Aims and Scope

The series aims at fostering the discussion on the complex relationships between physical landscapes, natural resources, and their modification by human land use in various environments of Asia. It is widely acknowledged that human-environment interactions become increasingly important in area studies and development research, taking into account regional differences as well as bio-physical, socioeconomic and cultural particularities.

The book series seeks to explore theoretic and conceptual reflection on dynamic human-environment systems applying advanced methodology and innovative research perspectives. The main themes of the series cover urban and rural landscapes in Asia. Examples include topics such as land and forest degradation, glaciers in Asia, mountain environments, dams in Asia, medical geography, vulnerability and mitigation strategies, natural hazards and risk management concepts, environmental change, impacts studies and consequences for local communities. The relevant themes of the series are mainly focused on geographical research perspectives of area studies, however there is scope for interdisciplinary contributions.

More information about this series at http://www.springer.com/series/8560

Huỳnh Anh Chi Thái

Livelihood Pathways of Indigenous People in Vietnam's Central Highlands

Exploring Land-Use Change

Springer

Huỳnh Anh Chi Thái
Human Geography, Institute of Geography
Heidelberg University
Heidelberg, Germany

ISSN 1879-7180 ISSN 1879-7199 (electronic)
Advances in Asian Human-Environmental Research
ISBN 978-3-319-71170-6 ISBN 978-3-319-71171-3 (eBook)
https://doi.org/10.1007/978-3-319-71171-3

Library of Congress Control Number: 2017960829

Printed on acid-free paper

This Springer imprint is published by Springer Nature
The registered company is Springer International Publishing AG
The registered company address is: Gewerbestrasse 11, 6330 Cham, Switzerland

Foreword

The book of Dr. Thái Huỳnh Anh Chi deals with the livelihood situation of minority people in the South Vietnamese highlands after the reunification of the country in 1975. As one of the many mountain areas in Southeast Asia, this region is influenced by a massive land-use change provoked by external factors such as coffee producers and also in-migration of "Kinh" people (ethnic Vietnamese, mainly from the coastal areas and north of the country).

In three different research areas, Dr. Anh Chi analyzes the actors of these changes, the possibilities and constraints of "making a living" for the indigenous people, and their resilience against pressures and activities for changing their complicated life situation.

The PhD thesis of Dr. Thái Huỳnh Anh Chi is one of the first empirical studies on this region, and the results can be generalized for other peripheral regions in Southeast Asia. Her work gives deep insights into village structures and their situation under the governance of a neoliberal economic but still communist country.

Geographisches Institut
Universität Heidelberg
Heidelberg, Germany

Hans Gebhardt

Preface

As a young person, who was born and grew up in the Central Highlands of Vietnam, I became familiar with the development of ethnic minority groups who lived adjacent to the city. The persistent gap in development between these indigenous communities and the Vietnamese group has been superficially blamed on the communities' backward thinking and high rate of illiteracy. In recent studies, in particular the research on development, the situation of minority groups has been examined more thoroughly. These communities appear to be victims of the transformation in the Central Highlands after the Vietnam War. They are named a vulnerable group, and the constraints they are coping with are considered as the vulnerability context.

Thus far, most research places indigenous people in a passive position, and little attention has been paid to their role as active agents in choosing their livelihoods. Therefore, I was motivated to examine their actor role and how certain constraints shape their actions of making a living. These pressing questions finally became the topic for my doctoral dissertation more than 3 years ago. I challenged myself by not adhering to the well-known theory of vulnerability, although it is a fundamental concept. Instead, I applied a hybrid concept in which vulnerability is accompanied by the actor approach and the resilience concept. It brought me to the discussions of the complex problems at three study sites in the Central Highlands (Kontum, Lak, and Lac Duong). Finally, one livelihood option, community-based tourism, was examined to see whether it could be a sustainable solution for the livelihood constraint of the indigenous community.

This work is expected to contribute to questions about the vulnerability of the ethnic minority groups in the Central Highlands of Vietnam, and possible findings for empirical cases are crucial to help decision-makers form effective policy interventions.

However, nothing could have been achieved without the following support and contributions.

I would like to give my sincerest thanks to my supervisor, Prof. Dr. Hans Gebhardt, for his mentorship and constructive guidance, as well as his patience and understanding. He gave me the opportunity to embark on this project and continued encouraging me during different phases. I appreciate the crucial comments and suggestions at the beginning of my work from Prof. Dr. Annika Mattissek and Prof. Dr. Paul Reuber. This study cannot be finished without a fieldtrip. Many thanks also go to Ms. Kim Jee Young and Mr. Nguyen Ngoc for their valuable advice to navigate my fieldwork and to my research assistant, Thanh Lan, for her tireless effort in helping me in the field. I wish to acknowledge Ho Chi Minh City University of Culture, Can Tho University, and state agencies at Kontum, Dak Lak, and Lam Dong Province for their support of my data collection and data analysis. I would like to give special thanks to communities in Kontum City, Lak District, and Lac Duong District for accepting me and being such willing participants in my research.

My thanks are extended to all staff members, the HiWis, and my fellow doctoral students in Prof. Dr. Gebhardt's group for their company, friendship, valuable comments, and sincere support. They are Dr. Klaus Sachs, Dr. Simon Runkel, Holger Köppe, Diana Griesinger, Schniepp Volker, Antonia Opelt, Warangkana Thawornwiriyatrakul (Pink), Jinliao He (Victor), Mehdi Ebadi, Guo Jie, Bonato Michela, Sok Sopheaktra (Tra), Azadeh Akbari, Juri Kim, and Peter Wongpan. My greatest appreciation and love go to my beloved family and my best friends for their unconditional love.

Finally, I am particularly grateful to the Katholischer Akademischer Ausländer-Dienst (KAAD) for their scholarship and the Kurt-Hiehle Stiftung Heidelberg and Fazit Stiftung for their contributions in making this research possible.

Heidelberg, Germany Huỳnh Anh Chi Thái

Contents

Abbreviations

CBT	Community-based tourism
GCF	Global Competitiveness Facility
GIS	Geographical information system
GOS	General Statistics Office
IPCC	The Intergovernmental Panel on Climate Change
JICA	Japan International Cooperation Agency
NEZ	New economic zone
NGO	Nongovernmental organization
PAR	Pressure and release
SFEs	State forest enterprises
SFs	State farms
SLF	Sustainable livelihoods framework
USAID	United States Agency for International Development
VHLSS	Vietnam Household Living Standard Survey

Chapter 1
Vulnerability Context: A Study on Livelihood Pathways of the Indigenous People

Abstract The socioeconomic transformation in Vietnam's Central Highlands has achieved good results in developing diverse cash crops that have a high output. However, this situation pushed indigenous inhabitants into the vulnerability context of livelihood constraint, which is caused by a lack of land and insufficient key assets. To examine the specific context of the indigenous people, this chapter forms a theoretical framework, which is constructed by three components of vulnerability (exposure, sensitivity, and resilience). More importantly, the actor concept is integrated in this framework to form the core connection of the three components and shed light on the active agent role of indigenous people. In addition, this chapter deals with the main methods, which combine spatial data and survey data.

Keywords Livelihood pathway · Indigenous people · Vulnerability group · Central Highlands

1.1 Background

Over the last half century, Vietnam's Central Highlands has undergone rapid environmental and socioeconomic transformations. The impact of these upheavals is strongly denoted by the following impressive numbers: "the population rose from 420,000 in 1926 to over 2.8 million by 1991 and then to more than 5.5 million in 2014" (Pamela McElwee 2008a; General statistical office 2015), and "the regional forest which had been very extensive until 1960 (covering up to 85% of the natural land) were lost at a rate of 30,400 ha per year between 1976 and 1990" (Rietbergen-McCracken et al. 2007, p. 52).

The events in Vietnam's Central Highlands are similar to those in other uplands of Southeast Asia and can generally be explained by the pressures of a developing economy and recovering postwar country. The Vietnamese uplands consist of the northern mountains and Central Highlands. The dynamic development has concentrated on the Central Highlands since it has become an agricultural frontier for people from all parts of the country (Hardy and Turner 2000, p.1). Resettlement processes and the promotion of cash crop cultivation have transformed Vietnam

H.A.C. Thái, *Livelihood Pathways of Indigenous People in Vietnam's Central Highlands*, Advances in Asian Human-Environmental Research,
https://doi.org/10.1007/978-3-319-71171-3_1

from "colonial possession to socialist state and to a current-day open market regime" (The World Bank 2009, p.2). Significant and positive changes have resulted. The agricultural sector has shifted from self-subsistence to cash crops with the commercialized production. In the 1990s, the coffee area expanded on a large scale, and this enabled Vietnam to surpass Colombia as the world's second largest coffee producer (Dang and Shively 2005, p.107).

However, there have been negative effects as well. Deforestation, degradation of soil and water resources, and unsustainable agricultural production practices appearing as side effects of the intensive agriculture have aroused concern, for example (Dang et al. 2001). An equally important issue is that the resettlement has completely changed the ethnic composition of the Central Highlands. These uplands had been a mosaic of 15 different indigenous groups and low numbers of the ethnic majority (called the Kinh). This population structure remained stable until 1975 (ITTO and IUCN 2005), but since then the indigenous groups have become minorities on their ancestral land. Hand in hand with rapid structural changes has come a growing disparity between indigenous groups and newcomers in terms of who reaps the benefits and opportunities. While the Kinh majority (and also the other migrating ethnic groups) have been the primary beneficiary, indigenous households have remained stuck in poverty and there is a persistent gap in welfare between these groups. The success story of the intensive agricultural development of the Central Highlands has not alleviated the severe poverty of its indigenous people. In this context, indigenous people are assessed as a vulnerable group (United States Agency for International Development 2008).

Research on transformations and vulnerable groups has developed in three different directions. The first research direction focuses on transitions in natural resources, in particularly land-use change (deforestation and conversion of forest lands into agricultural areas). This issue is a serious problem and has been traced by applying advanced technology and digital material such as satellite images and aerial photographs (Müller and Zeller 2002). Spatial data analysis offers valuable insights into processes of land-use change. Subsequently, the scholars have conducted many investigations to reveal the underlying causes of the dynamics of biophysical landscapes. The variables of land-use change can be either local, such as immigration and deforestation, or involve more distant factors, such as national policies (Leisz et al. 2009, p.237). In the case of Central Highlands, the scholarly consensus is that policies and market conditions are the main factors of change in land use. Many studies have focused on the milestone of "after 1975," as that was a period of dramatic changes in the Central Highlands. The change in land use was first triggered by national programs, such as the establishment of New Economic Zones (NEZs), State Farms (SFs), State Forest Enterprises (SFEs), and a mass organized migration to the Central Highlands to awaken this so-called "sleeping princess" area. Subsequently, the high market price of coffee sparked an influx of spontaneous migrants to these uplands to seek their fortunes in the commercial farming. This brought about a rapid expansion in coffee production in the 1990s.

In contrast to the region's success in promoting agriculture and developing the economy is the vulnerable situation of the indigenous people, which is the second research interest. Although indigenous groups do not always appear in the research due to their poverty and uncertainty, the transitions in their society, their political positions, and their cultural-social interactions are all potential sources of trouble. And left behind seems to be their common position in a number of ways. The general image of highlanders is one of low literacy, low income, disease, and lack of clean water (McElwee 2008a; The World Bank 2009). For more insight, scholars focus on the two main points of tensions in indigenes' lives, namely, land tenure and religion (Human Rights Watch 2002; McElwee 2001). These studies portray the indigenous community as "victimized"; their political and social situation and the effect on their lives are trenchantly assessed. The changes in these people's lives have led to a number of unanticipated results, most remarkably the conflicts related to land. Land grabbing and the resulting pressure on traditional cultivation have caused great bitterness among many indigenous communities (McElwee 2008a, b, p.201).

The third research dimension regarding the Central Highlands area is the livelihood dynamics and poverty of the Central Highlands. The considerable income gap between indigenous people and other groups, as well as the real depth and severity of poverty among indigenous communities, is a matter of some concerns. To understand the continued economic and social marginality of highlanders, many studies have been designed to explore the causal factors, ascribed to unequal access to land, education, credit, and other services (The World Bank 2009; Wells-Dang 2012). Subsequently, many projects have been implemented to consult and help indigenes to alleviate poverty and improve their lives.

However, most research places indigenous people in a passive position. From socioeconomic transformation to livelihood dynamics, much attention has been given to the external conditions and factors that appear to influence indigenous people (including political, social, and economic aspects). Even when discussing the drivers of poverty or the factors that contribute to reducing poverty, most considerations involve external factors. Little attention has been paid to the role of indigenous people as active agents of their communities. In order to highlight the ability of communities to mobilize and cope with their vulnerability contexts, it is necessary to place more emphasis on the indigenous people as actors and on how certain constraints shape their actions. A better understanding of the complex interactions between transformations and the lives of indigenous people, in particular their livelihoods, is crucial to help decision-makers form effective policy interventions.

In this present study, the complex problem of the highland minorities' livelihood situations will be disaggregated, and the agentive behavior of indigenous people can be investigated to answer the following questions:

- To what extent do land-use change and socioeconomic transitions influence the livelihood dynamics of the indigenous people in the Central Highlands of Vietnam?
- What are the other internal factors driving the livelihood pathways of indigenous households?
- To what extent can indigenous people take an agentive role in their own livelihoods?

1.2 Key Research Concepts: Vulnerability, Actor, Livelihood Resilience

The empirical study presented here focuses on the livelihood situation of the minorities in Vietnams' mountainous area, on the different actors involved, and on the dynamic changes that have occurred during the last few decades. The rational actions of actors are analyzed in relation to their vulnerability context, and this, therefore, forms the core of the theoretical framework and methodologies applied in this study.

As Bohle (2001) stated, vulnerability is being shaped by a complex and interactive system of "external" and "internal" contributing factors. The external factors relate to dynamic conditions comprising both environmental and social elements to which the internal factors are exposed. However, internal factors do not only appear as passive victims. Thus, the actor concept is deployed to demonstrate the significant capabilities of the internal system to be constantly flexible and responsive. And finally, community-based tourism (CBT) is explored as a livelihood strategy to validate the local people's resilience capacity. Vulnerability is viewed as a general concept that embraces the others, and CBT is the smallest unit of the conceptual framework. Meanwhile, the actor approach is placed at the center to allow for an analysis from dynamic conditions to a certain livelihood. In other words, under the actor-oriented approach, this study is conducted from the macro-level of the vulnerability context as the backdrop to the micro-activity of CBT.

Three aspects above are integrated to form the conceptual frame of this research. Their relationship can be described by this below overlapping design (Fig. 1.1).

1.2.1 Vulnerability and Human-Environment Interaction

The concept of vulnerability has been covered in a large body of academic literature. In the last decades since some initial research on vulnerability appeared, the concept of vulnerability has been more widely studied from various perspectives. A comprehensive analysis of vulnerability and relating notions have been summarized by a number of authors (Cutter 1996; Prowse 2003; Thywissen 2004; Birkmann 2006). However, they have not yet reached consensus regarding this term.

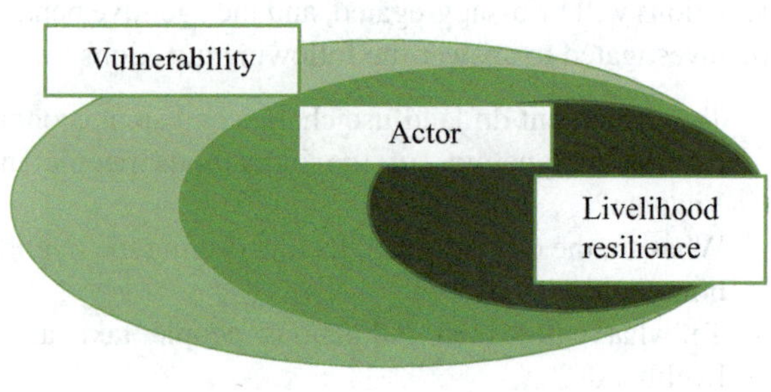

Fig. 1.1 Major research areas relevant to study (*Source*: Huỳnh Anh Chi Thái)

Discrepancies in the meaning of vulnerability arise from different epistemological orientations (e.g., geography, physics, economics, environment, sociology), the considerable variation in the choice of hazards themselves, and the regions chosen for examination (Cutter 1996). The heterogeneous meaning is subsequently accompanied by different methodological practices.

Theoretical Concept: Biophysical and Social Vulnerability

The plurality of definitions for vulnerability has led to various and intensive conceptual works (e.g., Wolf et al. 2013). These different vulnerability concepts[1] address various perspectives, such as vulnerability situations, vulnerability factors, vulnerability sides, and vulnerability dimensions. A frame of vulnerability factors is deployed in this study to get an overarching context generated by different driving forces and processes.

Several researchers have distinguished the vulnerability factors as being biophysical (or natural) and social (or socioeconomic) factors. Despite the wide range of collective research on this concept and the same categories being used, these two factors still mean different things to different authors.[2] In Füssel's research (2007), he attempts to reconcile the various conceptualization of vulnerability and then assumes that "none of them is comprehensive enough to consistently integrate the others." He then points out that this confusion derives from the failure to distinguish between two largely independent dimensions of vulnerability factors: sphere (or scale) and knowledge domain. Due to the confused lexicon of meanings, the discrepancies in the terminologies of vulnerability (Cutter 1996), and the lack of comprehensiveness to consistently integrate each concept (and its sub-concepts) (Füssel 2007), it is necessary to delineate the spheres and clarify the positions when discussing the chosen concept.

Biophysical Vulnerability – Natural Hazard

Two major research traditions in vulnerability are the analysis of vulnerability as a lack of entitlements (Sen 1981) and the analysis of vulnerability to natural hazards (Burton 1978 and 1993, cited by W. Adger 2006). While the first entitlement approach analyzes vulnerability to famine (Watts and Bohle 1993), the second

[1] Conceptual works usually bifurcate into two opposing approaches: "biophysical" and "social", "outcome" and "context", "current" and "future" (Adger 2006; Wolf et al. 2013), "pre-existing condition" and "tempered response" (Cutter 1996), "internal side" and "external side" (Bohle 2001; Chambers 2006), etc. However, it is not the purpose of this research to review and gain insight into each conceptual frame. For a summary of approaches to vulnerability, the reader is directed to W. Adger 2006 and Füssel 2007.

[2] Klein and Nicholls (1999) regard "natural vulnerability" as one of the determinants of "socioeconomic vulnerability"; Brooks (2003), in contrast, regards "social vulnerability" as one of the determinants of biophysical vulnerability. And in a different point of view, Cutter (1996) sees "biophysical" and "social" vulnerability as parallel processes that exist independently.

antecedent researches most prominently in natural hazards, which limits to the impacts of natural hazards on society (Kreimer and Arnold 2000, cited by Fünfgeld 2007). The vulnerability literature refers to the fields of climate change, environment, and natural hazards. In many research articles, the term hazard especially refers to physical manifestations of climate variability or changes that relate to natural phenomenon such as droughts, floods, storms, and long-term changes in the mean values of climatic variables (Cutter 1996; Brooks 2003). Biophysical vulnerability is then featured by the magnitude, duration, impact, frequency, and rapidity of onset.

This natural hazards approach views vulnerability as a pre-existing condition from which a community in hazardous zones suffers (Cutter 1996; Tahmasebi 2012). Biophysical vulnerability is also used in the "hazard-oriented" approach. In this classification, authors suggest that biophysical vulnerability represents a "state of nature," existing independently from a human perspective (Chazal 2010). Moreover, Brooks (2003) suggests that the term "biophysical vulnerability" not only refers to the sphere of "physical component" associated with the "nature of the hazard" but also regards it as first-order physical impacts. This opinion results in downplaying and neglecting the role of human systems. In this line, it will be examined more in terms of human exposure to hazard than people's ability to cope with hazards (ibid).

However, when primarily focusing on biophysical vulnerability factors based on the hazard-oriented approach, this concept is hardly applicable to environmentally induced causes of human vulnerability, which is simultaneously determined by people's agent-oriented and socioeconomic situations (Tahmasebi 2012).

Social Vulnerability – Human-Induced Stresses

As opposed to the biophysical approach, social vulnerability has been analyzed from a people-oriented approach (Tahmasebi 2012). This approach is termed differently by different scholars, such as "social response" (Cutter 1996), "social vulnerability" (Brooks 2003), "human ecology" (Adger 2006), or "political economy" (Füssel 2007). In this research, we propose to use "social vulnerability" in the opposite meaning of the biophysical vulnerability discussed above.

While biophysical vulnerability is related to external nature properties, social vulnerability is viewed as an inherent property that arises from its internal characteristics. The social space of vulnerability is determined by the political, economic, and institutional capabilities of people in specific places at specific times (Watts and Bohle 1993). In more detail, it refers widely to the social realm of institutions, distribution of power, social status, cultural practices, and other social characteristics (Adger 2006; Füssel 2007) and is addressed by determinant factors such as poverty and inequality, marginalization, food entitlements, and access to resources (Brooks 2003; Adger 2006). This approach is primarily applied in the social research fields that are concerned with identifying the most vulnerable members of society (Brooks 2003). The role of social factors in vulnerability outcome is known as "the term of entitlement" – one of two antecedents in vulnerability research which is judged to be successful in highlighting social differentiation and underplayed ecological or physical risk (Adger 2006).

Since social vulnerability encompasses all the properties of a system exposed to hazards, it is viewed as the factors that determine the outcome of a hazardous event (Brooks 2003). From this point of view, the exposure to hazard will depend on where population choose to (or are forced to) live and the characteristics of communities and their livelihoods. It not only focuses on human exposure but also looks at people's ability to cope with hazards. The social system increases or reduces the damage resulting from the biophysical side. When discussing the outcome of social vulnerability, it should focus on two processes: the response of systems (households, communities) that may affect the livelihood system and the transformations of this response to a changing political economy (Fünfgeld 2007).

However, research on social vulnerability does not look specifically at the social impacts of natural hazards. In the present study, social vulnerability is described not only as the outcome of a hazard but also the context of this hazard event. The hazard results from the interaction of the social vulnerability and natural hazard. It is too risky to place too much emphasis on natural hazards and separate them from the social framework (Wisner et al. 2003). As such, it is necessary to set a balance in assessing the causes of vulnerability to both natural processes and socioeconomic and political factors.

Therefore, while the biophysical factors are obviously viewed as external factors, the sphere of social vulnerability should be considered in terms of both internal and external factors. The delineation between external and internal vulnerability factors depends on the scope of the vulnerability assessment (in this study, the scope of the vulnerability assessment is the communities of indigenous people in Vietnam's Central Highlands) and will come back later in more detail in the discussion about the concept of actor.

Integration

Disasters are viewed as a result of complex interactions between a potentially damaging physical event and the vulnerability of society due to human behavior (Birkmann 2006b). In this line, vulnerability can be understood in a holistic manner, including both natural and social systems. When discussing the evolution of approaches to vulnerability, Adger (2006) states that previous research on vulnerability is now being translated into research on the vulnerability of social and physical systems in a more integrated manner. However, researchers use different approaches to translate this.

In their comprehensive research that gives equal weight to both social and biophysical vulnerability, Blaikie and his colleague successfully synthesize the two traditions of hazards research in their hazard model "Pressure and Release" (PAR) (Wisner et al. 2003). In this model, biophysical hazards are viewed as one characteristic of vulnerability, with the other pressure resulting from social factors

(Adger 2006). This human-nature framework captures the context of vulnerability and determines the outcome of vulnerability by social differentiation.

In addition, several researchers have presented spatial context or geographic space as a unit when examining vulnerability context and have proposed the "hazard of place" model (Cutter 1996). However, this model downplays the internal interactions of biophysical and social factors and evaluates these factors in parallel processes. For the specific purpose of this study, the author applies an approach which not only covers aspects of biophysical stress and impacts of policy change, social transformation, and economic fluctuation but an approach that also pays equal attention to understanding that vulnerability lies in the interaction between social dynamics and biophysical processes in a specific space. This orientation is supported by the "place-based" model of Turner et al. (2003). The term "place-based" implies spatially distinctive elements of social and biophysical conditions or coupled human-environment systems. A particular strength of place-based analysis is its potential to achieve a collaborative assessment. However, synthesizing these juxtaposed approaches into a coherent concept can be problematic.

This integrated concept is inspiring and useful for the subsequent empirical study; also it is suitable for geographical research with an epistemological perspective and disciplinary focus on the interface between social and natural spheres.

Dimensions and Components of the Vulnerability Concept: Choices and Applications

Due to its wide applicability in various fields, the definition of vulnerability, its components, and meaning are growing increasingly. Vulnerability is often conceptualized as constituting a range of components such as exposure, sensitivity, susceptibility, resilience, coping, and adapting (Adger 2006; Birkmann et al. 2013; Bohle 2001; Robert Chambers 2006; Moser 1998; Prowse 2003; Turner et al. 2003). In spite of the differences in terms, the meanings of these components overlap each other. In this study, some components of vulnerability will be chosen to present.

Exposure

From the initial research on vulnerability, Chambers (2006) states that "vulnerability refers to exposure to contingencies and stress and difficulty in coping with them." According to Birkmann et al. (2013), "exposure of a society or system to a hazard or stressors" is the key factor of the common vulnerability framework. In the definition of vulnerability used by the IPCC (Intergovernmental Panel on Climate Change) – also one of the most popular ones (Adger 2006) – vulnerability is understood as a function of exposure, sensitivity, and adaptive capacities.

"Exposure," as the first component of vulnerability, emerges from the research on and practice to combat risks and hazards and has a limited focus on perturbations and stressors that expose the systems. In the past, a hazard was often understood as a natural event and exposure subsequently referred to the natural impacts that systems experience and cope. However, in Watt and Bohle's conceptual framework, exposure is defined in terms of external socioeconomic and sociopolitical conditions (Watts and Bohle 1993). In the same view, Birkman argues that the exposure of a system can only be adequately understood if the coupled human-environmental systems and interactions are addressed (Birkmann et al. 2013).

When considering vulnerability from both sides (biophysical and social sides), it is useful to view exposure in light of these processes. At the core of the definition of vulnerability, exposure is defined as follows: "Exposure describes the extent to which a unit of assessment falls within the geographical range of a hazard event. Exposure extends to fixed physical attributes of social systems but also human systems that are spatially bound to specific resources and practices that may also be exposed. Exposure is then qualified in terms of spatial and temporal patterns" (Birkmann et al. 2013).

However, not all previous research uses a specific assessment method or a predefined list of indicators for exposure. Exposure to the system in Vietnam's Central Highlands is examined by the linked processes of land-use change and socioeconomic transformations.

Sensitivity

Turner et al. (2003, p.8074) suggest that "[v]ulnerability is registered not by exposure to hazards alone but also resides in the sensitivity and resilience of the system experiencing such hazards." In this line, Bohle (2001) argues that while the external side of vulnerability exposure has been discussed elsewhere, the internal side of coping has so far been widely neglected. Sensitivity is covered in close connection with exposure and refers to the characteristic of how the internal system reacts to external phenomena.

However, similar to the other dimensions, it is defined in various ways. In the IPCC's fourth assessment report, sensitivity is defined with reference to climate change as "the degree to which a system is affected, either adversely or beneficially. This effect may be direct or indirect" (Parry et al. 2007). Meanwhile, Moser views sensitivity as the internal capability of a system and emphasizes the interaction between the external hazard and the internal capability in the sensitivity/resilience model of vulnerability (Moser 1998). In this case, sensitivity is interpreted as the magnitude of a system's response to an external event.

Since the system under exposure in this study refers to the local people, the definition of sensitivity denotes the characteristics which make them susceptible to the impact of the external human-environment system. In this meaning, the sensitivity concept can be viewed in light of the indicators relating to the distribution of power, social institutions, specific cultural background, cultural practices, and other characteristics of social groups (Füssel 2007).

Resilience

The last dimension of vulnerability is resilience. The resilience dimension is used in vulnerability analysis to distinguish between the twofold concept of vulnerability: exposing and coping. Bohle states that vulnerability consists of both external and internal sides. While the external side of vulnerability exposure has been widely discussed, the internal dimension, which focuses on coping and action to overcome or at least mitigate the negative effects of economic and ecological change, has been widely neglected (Bohle 2001).

Moser states that resilience can be viewed as the ease and rapidity of a system's recovery from stress, its responsiveness to risks faced during such negative changes (Moser 1998). The resilience of a system is often evaluated in terms of the amount of change a given system can undergo and still remain within the set of natural or desirable states. The resilience concept has been applied in a variety of interdisciplinary research focused on coupled human-environment system (Turner et al. 2003). In order to denote this concept, different terms are used, such as "response capacity," "coping capacity," and "adaptation capacity" (Füssel 2007). Resilience is attributed to two core components: coping and adaption (Birkmann et al. 2013). Coping signifies the short-term actions available to those at risk and mainly deals with the conservation and protection of the current system and institutional settings, while adaptation denotes a longer-term and constantly unfolding process of learning, experimentation, and change that feeds into vulnerability (Birkmann et al. 2013).

Since this study focuses on the livelihood dynamics, it is necessary to pay more attention to the adaption capacity (than to the coping capacity) which signifies the long-term change in a system and the outcomes of learning – planned and spontaneous and pre- and post-disaster (Pelling 2010). To sum up, vulnerability is denoted based on the chosen dimensions of "exposure," "sensitivity," and "resilience." The framework below based on those three key dimensions provides an integrative and holistic tool to assess the vulnerability context in Vietnam's Central Highlands (Fig. 1.2).

However, it is also important to note the similar and overlapping meanings of each component. While Watts and Bohle use "exposure," "capacity," and "potentiality" to define the space of vulnerability (Watts and Bohle 1993), Moser utilizes the concepts of "sensitivity" and "resilience" to design the two-step model of vulnerability (Moser 1998). Although the terms are different, there are key links and their meanings are similar. The "sensitivity" in Moser's model brings together both exposure and capacity aspects, while "resilience" represents capacity and potentiality aspect (Prowse 2003).

Figure 1.3 shows that both the sensitivity and resilience dimensions of the vulnerability concept include capacity. That is why Moser emphasizes assets and capabilities as a determinant of vulnerability (Prowse 2003). In this line, the role of capacity and characteristics of system will be clarified in the next section, when we discuss the concept of actor.

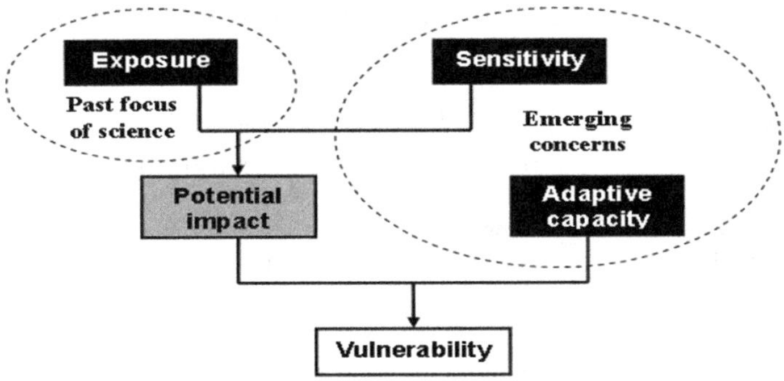

Fig. 1.2 Dimension of vulnerability concept (*Source*: Smith and Lynam 2010)

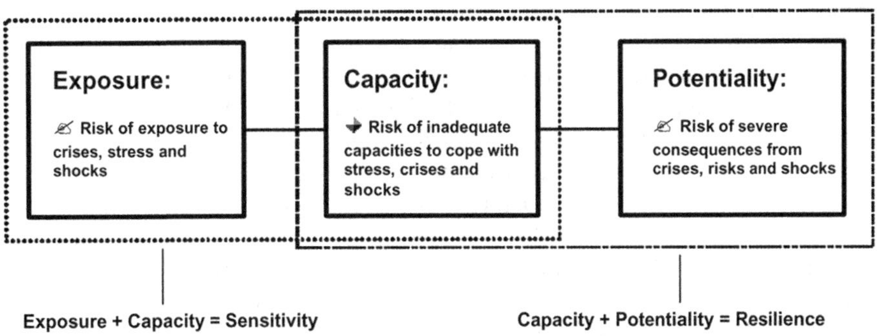

Fig. 1.3 Component and overlapping terminology in vulnerability research (*Source*: Prowse 2003)

1.2.2 The Actor-Oriented Approach and Internal-External Factors of Vulnerability

The couple system of natural hazards and human-induced stresses is just one side of the "double structure of vulnerability" (Bohle 2001). Bohle has raised this term to point out the differences between the external and internal sides of vulnerability. In order to integrate these above dimensions and components and reach a coherent analysis, the sphere of vulnerability as internal or external sides needs to be differentiated. To determine whether specific factors are internal or external, we first need to determine the scope of the vulnerability assessment (Füssel 2007). For example, according to the "dimension sphere" and "knowledge domain," Füssel classifies the vulnerability factors into four categories described in following figure. These categories can be alternatively classified or broken down further in regard to the aim of study and the specific assessment (Table 1.1).

Table 1.1 Four categories of vulnerability factors classified according to the dimension sphere and knowledge domain

	Domain	
Sphere	Socioeconomic	Biophysical
Internal	Household income, social networks, access to information	Topography, environmental conditions, land cover
External	National policies, international aid, economic globalization	Severe storms, earthquakes, sea-level change

Source: Füssel (2007)

Since the analysis of vulnerability not only involves identifying threats but also the response that individuals, households, or communities can mobilize to cope with hardships (Moser 1998), it is also necessary to include the values held by the actors as active agents of change. However, it is a challenge to conceptualize this actor concept into a research framework of vulnerability. As such, this study raises the question to what extend the actor can influence their vulnerability context? In light of these considerations, "actor" is used as the scope in the vulnerability concept and is placed at the center of theoretical foundation, which is used to precisely delineate the sphere of internal and external factors.

The actor-oriented model was developed based on two complementary theoretical discourses that both discuss the relation between the structure and the actor theories: one stemming from Bourdieu's theory of practice (Bourdieu 1977) and the other proposed by Gidden – the "structuration theory" (Giddens 1984). The key concepts of "theory of practice," which are habitus, practice, capital, and field, provide the pattern to interpret actions and strategies from the perspective of the relationship between actor and social structure (Wiesmann et al. 2011). According to Parker, habitus is a system of acquired, learned, and lasting dispositions to perceive, think, and act in certain ways (Parker 2010). It then triggers the practice and sets off actions or strategy. Habitus is shaped based on four main types of capital: economic, cultural, symbolic, and social (Wiesmann et al. 2011). The distribution of these key capitals, their interrelation, and their convertibility determine the power relations between actors, which are signified through the last concept of field. Field is denoted as a "set of social relations and a system of social positions" in which the strategies and outcomes of actions will be decided and by which the actions of actor can be constrained or supported (ibid: 237). Besides, Giddens's theory presents another fold of this social-actor interaction: an actor is characterized as being conscious and capable of having reasonable and intentional actions which can significantly influence social structures (Giddens 1984). In this line, the structuration theory places more emphasis on the actors' influence.

Using the actor concept in vulnerability research helps to examine local actors in an active role. From being passive and incapable of influencing their vulnerability context, an actor can shift to being active in their perception, valuation, interpretation, and anticipation. Moreover, the actor-oriented model allows for consideration of the dynamics of change and the dynamics between the context and the actor (Wiesmann et al. 2011). Thereby, it overcomes the vague and static nature of some concepts that emphasize the structures rather than actors. According to Wiesmann, the actor concept consists of "four nested and interlinked components," each representing a core conceptual element: meanings of action, means and assets, activities and practices, and institution (ibid).

Employing the actor concept in the vulnerability assessment helps to delineate the external and internal sides and then arranges components and factors into these two sides. This allows us to create a coherent relationship among the dimensions and integrate them into a concept framework. The external side is understood as "risk, shock, and stress to which an individual or household is subject" (Chambers 2006) and refers mainly to the structural dimensions of vulnerability (Bohle 2001). Structure, in this sense, not only refers to the natural hazard or macroscale events but also to changes that occur locally, in the proximity of the social unit or system (Fünfgeld 2007). In other words, since the actors in this research are the indigenous people, the external side is denoted by the exposure from biophysical, social factors (excluding the concerned actor – the indigenous), and the integration between biophysical and social factors as well. In contrast, the internal side refers to the issues within this system (the internal biophysical vulnerability is thereby not concerned). It is thus regarded to indigenous people's characteristics and their defenselessness that impinges on their individual or joint ability to cope with stresses and will be examined as an aspect of sensitivity and resilience. Since vulnerability comprises both an internal and external side in this study, the social vulnerability concept embraces both sides. The author divides these into the social stresses and the ability of individual's society to cope with exposure. The actor concept, which theoretically refers to the individual, is used in this study to speak about indigenous communities as active agents. In reality, institutions within indigenous community sometimes no longer exist or no longer play a strong role; the role as an active agent of community is thus determined by individual households instead.

The categories of vulnerability dimensions and their relationships discussed above are modified and illustrated in the following Fig. 1.4.

1.2.3 Resilience Livelihood: An Asset Approach and Livelihood Pathways of Indigenous Farmers

The empirical vulnerability context in this study relates to the constraint in land use and the socioeconomic issues that act as livelihood stresses. The present study does not conceptualize livelihood as a specific vulnerability but does emphasize livelihood stress as one of the main pressures on the indigenous people. The dynamics of their livelihood transformation are examined as responses of households to the constraints they are faced with. Therefore, the livelihood transformation will be considered as an integral part of the resilience dimension when it is included in the conceptual framework.

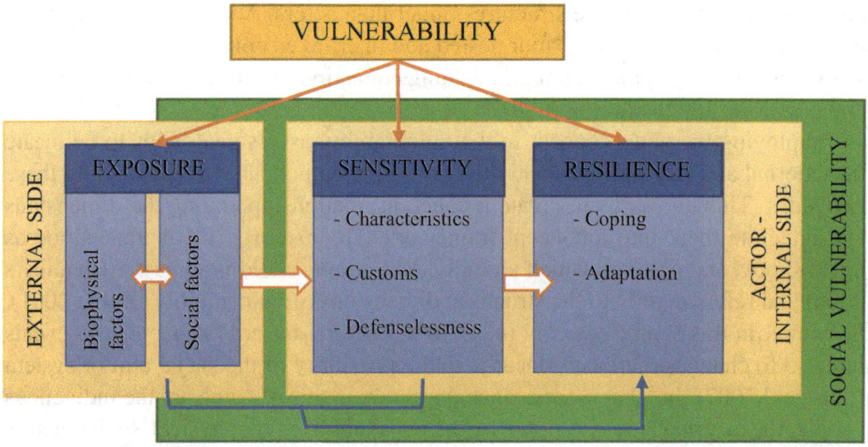

Fig. 1.4 Dimensions and components of the vulnerability concept (*Source*: Huỳnh Anh Chi Thái)

In a recent research, Speranza and her colleagues stated that: "Livelihoods thinking should be enhanced in relation to sustainability through incorporating the concept of resilience" (Ifejika Speranza et al. 2014, p.111). Moreover, according to Van Dillen (2004), vulnerability and livelihood can be viewed as two sides of the same coin: one refers to people's exposure to livelihood risk and the other to their capacity to cope with risk. However, the vulnerability concept places more emphasis on the external side from the macro-perspective, while livelihood emphasizes the internal side. To compromise these arguments and adopt them to the conceptual framework, livelihood dynamics will be examined as a component of the internal side and addressed as one aspect of resilience. However, as this component of the internal side of vulnerability is not easily viewed from a macro-perspective, Dillen suggests exploring this dimension from a micro-point of view (ibid). Therefore, one of the livelihood options will be chosen to assess the extent to which indigenous communities could adapt and overcome the constraints they face and to answer the question whether it is a good option to help indigenous people cope with their livelihood pressures.

A livelihood approach is formed by three main components: the livelihood context; the transformation and processes of aspects such as institutions, organizations, policies, and legislation that affect livelihood strategies; and the actors' assets. (Ifejika Speranza et al. 2014). The livelihood context and its processes are viewed as the exposure dimension of vulnerability which mediates livelihood strategies. Livelihood assets are different resources that people are entitled to mobilize to achieve their goals. Recently, the concept of asset has been used frequently when examining the internal dimension of vulnerability from a micro-perspective (Moser 1998; The Department for International Development 1999; Bohle 2001; Van Dillen 2004). Bohle argues: "The more assets people control, the less vulnerable they are and the greater are their capacities to successfully cope with risks, stress and shocks" (Bohle 2001, p.3). Meanwhile, Moser states that the means of resistance are the assets and entitlements that individuals, households, or communities can mobilize and manage to face with hardship (Moser 1998).

Livelihood resilience is then characterized by actors' assets and strategies to maintain and enhance assets (Ifejika Speranza et al. 2014). The strength of this asset concept is the ability to connect actors to their complex socioeconomic and natural environments (Van Dillen 2004). Exposure addressed from human-nature interaction using a social-ecological lens is seen as the root of vulnerability, whereas the asset perspective explains the differentiations of resilience in a local context from a higher-resolution view.

An actor is recognized as a manager of complex portfolios in which assets are interrelated, complementary, and substitutable (Moser 1998; The Department for International Development 1999; Van Dillen 2004). The sustainable livelihoods framework (SLF) proposed by the British Department for International Development, which focuses on human being as real agents of change, has been popular in development research over the past decades. The SLF structures the three components of livelihood as mentioned above. Compatible to other components in the internal side of the vulnerability concept, the livelihood approach puts indigenous people, along with their assets, choices, and actions, at the center of development. Assets are considered the core of the livelihood concept (Fig. 1.5).

The livelihood approach seeks to gain a realistic understanding of people's strengths (assets or capital endowments) and the way they strive to translate these into positive livelihood outcomes. It is assumed that no single asset can sufficiently achieve the livelihood outcome, particularly in the vulnerability context. This approach requires the mobilization of a range of assets to ensure the livelihood outcome (The Department for International Development 1999). This study empirically considers these five capitals: natural, physical, economic, human, and social assets. And the scope of capital assets is determined as following:

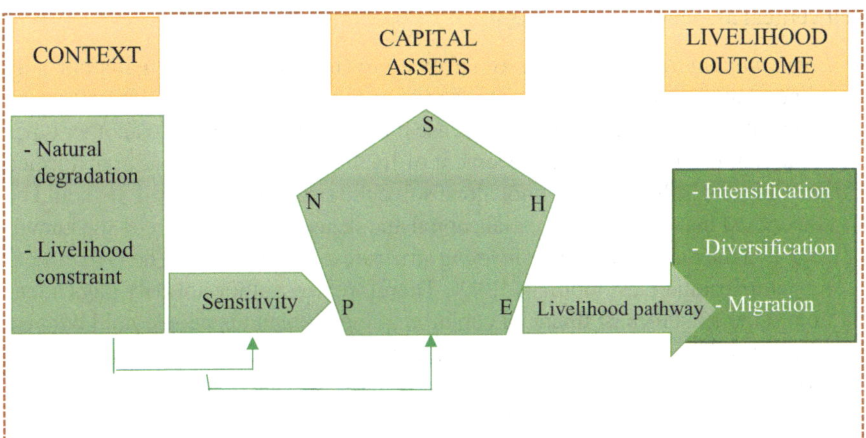

Fig. 1.5 Livelihood framework. Key: *S* Social capital, *H* Human capital, *E* Economical capital, *P* Political capital, *N* Natural capital (Modified from DfID 1999)

- *Human capital* concentrates on the agency of human beings. It comprises the skills and knowledge of an individual (Van Dillen 2004) and refers to an individual's ability to labor and therefore represents good health and physical fitness. In the context of this study, where traditional knowledge plays an important role in people's daily and economic activities, it is worth weighting more on the accumulated experiences. Human capital is a factor of the amount and quality of labor available and varies according to household size, skill levels, leadership potential, health status, etc. (The Department for International Development 1999).

In the framework of livelihood, human capital has an intrinsic value which is necessary and required to activate the other assets. However, human capital and its value are, in turn, affected by the socioeconomic context. This means it is necessary to view human capital through a macro-perspective.

- *Social capital* has been identified in a variety of discipline contexts and by different terms such as social organizations, behavioral norms, and social networks (Van Dillen 2004). According to Ellis, social capital refers to an individual's or household's major networks, relationships of trust, and wider institutions upon which people draw in pursuit of secure livelihoods (Ellis 1999). Meanwhile, Moser states that social capital refers to reciprocity within communities and households based on trust derived from social ties (Moser 1998).

This concept has been discussed widely. According to Coleman (1988), social capital includes vertical as well as horizontal associations. While the horizontal associations represent the relationship between people who have an effect on the productivity of the community, the vertical associations are characterized by hierarchical relationships and unequal distribution of power. This concept provides a comprehensive perspective to analyze social capital in this study. We look at formal and informal networks to which the indigenous people belong, from both horizontal and vertical associations.

- *Economic capital,* which is also narrowed to financial capital, entails not only money but also access to formal loans or personal credit (Tran 2012). Though financial capital is important in the sense that it can be converted to another types of capital, used for direct achievement of livelihood outcomes, and transformed into political influence, it cannot alone solve the stress on livelihood. It should be considered that access to financial capital and its usage is governed by the knowledge and the existence of transforming structures and processes (The Department for International Development 1999). Therefore, the author not only pays attention to the existence of financial capital that the indigenous people hold but also looks at the types of financial organization services available, the services they provide, and their conditions and access to these services.
- *Physical capital* comprises the basic infrastructure and producer goods that individuals/households need to support their livelihoods (The Department for International Development 1999). It also represents the totality of an individual's

material possessions and investments (Johson 1997, Stewart 1998 cited by Van Dillen 2004). It may thus include road networks, electricity, medical clinics and hospital, schools and markets, housing, land, irrigation, etc. In some cases, the lack of physical capital is considered to be a core dimension of poverty. Insufficient or inappropriate producer goods also limit the mobilization of other capitals and constrain people's productive capacity (The Department for International Development 1999).

Given the context of indigenous people who heavily depend on cultivation and live in remote areas far from social facilities (schools, hospitals, markets), the analysis of physical capital focuses on road networks, communications, and the infrastructure supporting cultivation in general.

- *Natural capital* is the term used for the natural resource stocks useful for livelihoods. There is a range of resources that make up natural capital. The natural capital is important to those whose livelihoods depend on resource-based activities (The Department for International Development 1999). In this line, the vulnerability context used in this study closely reflects this natural capital, since the indigenous people heavily depend on the forest and land for their traditional economic activities. The degradation of natural resources and the constraint in land-use change place strain on their livelihood and shape their vulnerability.

To assess natural capital, a range of information is required such as the existence of different types of natural assets, ability to access and combine them, and the quality and variation of those assets over time (The Department for International Development 1999). While natural capital ranges from intangible public goods to divisible assets, this study pays particular attention to land use, property rights, and access to forest, which has direct effect on the indigenous people's livelihoods.

1.2.4 Vulnerability in Vietnam's Central Highlands

The present study was conducted in Vietnam's Central Highlands (*Tây Nguyên* in Vietnamese). This is the last exploited land in the country. Since the country's reunification in 1975, a series of institutional and policy reforms have been implemented in this area. As a result, the Central Highlands have achieved the high agricultural output and the diversification in agricultural production. Along with these achievements, over the past four decades, this area has undergone rapid social, economic, and environmental changes. Deforestation, natural depletion, and poverty are key constraints that make up the vulnerability context for the local people, in particular, the indigenous people: the vulnerable groups.

Vulnerability Situations and the Actor Role of Vulnerable Groups

Our theoretical framework includes elements of different theories of vulnerability, actor, and resilience. It has been modified based on concept mainly developed at the South Asia Institute in Heidelberg and adapted to the context of the Central Highlands in Vietnam.

In this research, the indigenous people are viewed as active agents. Their characteristics and their actions demonstrate not only their vulnerability in being exposed but also their resilience to external socioeconomic and natural pressures. Linguistically, the indigenous people in this study are composed of numerous ethnic groups from the Austronesian (Malayo-Polynesian) and Austroasiatic (Mon-Khmer) language families. Historically, this mountainous area is their ancestors' land and has never before been populated by the ethnic Vietnamese, the Kinh. However, after a range of changes in the indigenous people's socioeconomic conditions, they have become minorities in their own land. Subsequently, their socioeconomic status has changed. Nowadays, the indigenous people are characterized by poverty, unemployment, and illiteracy. Previously, their income was based mainly on agriculture and they partly relied on the natural forest for products such as food, fodder, medicines, and construction material (Sunderlin and Ba 2005). Agriculture and forestry, thus, constituted the basis of the indigenous economy. The change in land use and the change to their traditional agricultural system have put their livelihood under great stress. While the status of land-use change is clear, the issues relating to land pressure and strategies to cope and overcome this problem remain a serious concern among scholars and resource managers in the region (Espaldon et al. 2004; Leisz et al. 2009). In this research, land-use change is viewed as a cause of vulnerability and is used to learn about the sensitivity and resilience aspects of the local people.

However, the vulnerability context should not be superficially viewed as a natural vulnerability only but should also be examined in light of socioeconomic pressures. This research looks for a historical root cause of the changes in natural resources. The vulnerability context results from the interaction between these natural and social factors.

Sensitivity and the Livelihood Pathways of the Indigenous People

Sensitivity in this study refers to the characteristics of indigenous people that make them susceptible to external factors. Normally, households with a vulnerable livelihood system practice various kinds of activities, also called their livelihood portfolio, which allows them to cope with external stresses. The process that the indigenous people follow to make livelihood decisions should not be characterized as a preset goal which is then attained as a livelihood strategy. Instead, in the context of risk and stress, there must be a livelihood pathway which allows for iterative process, guided by past decisions that contribute to their present mental attitude (Tran 2012).

The livelihood pathways of indigenous people in the Central Highlands are shaped by their choices and reflect their characteristics, their nature, and specific cultural background. Therefore, clarifying features of their livelihood pathways may partly reveal their sensitivity, which is determined by their intrinsic nature. The comparison between the status of indigenous people and that of the other groups who share the same external resources will shed light to the question: How the various social actors gain access to and control over such resources (Leach et al. 1999)? In order to address this aspect, we draw on the entitlement concept. Although the first entitlement approach developed by Sen (1981) narrowed entitlement down to the access and control over natural resources when examining with poverty and famines, an extended entitlement approach is applied in this study.

In his work, Sen argues that famines are not caused by scarce food resources but are a result of the institutional failure in distributing the available food. This process is generalized when he uses the term "endowment" and "entitlement." While endowments refer to resources owned by the actors, entitlements refer to the actors' access to resources (ibid). The entitlement set can be acquired by converting endowments through exchange entitlement mapping (e.g., production, trade, social security provision, and subsidies) (Devereux 2001).

This approach contributes to shedding light on the situation of the indigenous people: their vulnerability context is not only due to the lack of endowment (land) but also on their ability to access the resources and mobilize them. Their entitlement mapping, which is partly determined by their characteristics, can enhance or inhibit the access. This then shapes their livelihood pathways.

Community-Based Tourism as a Livelihood Resilience

Community-based tourism is discussed in this study as a particular livelihood option to examine the resilience of indigenous communities. This livelihood option represents the transformation that local people go through when learning to cope with constraints. And the level of resilience will be assessed by looking at how communities mobilize assets to make a living.

Diversification and the Role of CBT

Rural residents struggling to make a living can construct their livelihoods using three main strategies: agricultural intensification, livelihood diversification, and migration (Hussein and Nelson 1999). Since they have neither enough assets nor the capabilities to access them, specializing in agriculture alone may be insufficient to cope with the stresses they face. Instead, rural livelihood should be diversified to encompass a portfolio of different resources and activities (Niehof 2004). According to Hussein and Nelson (1999), livelihood diversification includes both on- and off-farm activities that are undertaken to generate income in addition to the income generated from the main household agricultural activities.

This study pays more attention to off-farm income as a way to multiply livelihood. Of them, tourism is considered an effective way of making a living, as it brings in extra income, and is not affected by land constraint problems. Moreover, community-based tourism (CBT) promises to be a tool that can support sustainable development. CBT not only provides direct benefits, such as profits and jobs, but also has the huge potential for diversifying the local economy, as well as a range of indirect positive impacts such as conserving natural resources and preserving cultural values (Hamzah and Khalifah 2009). Another significant feature of this tourism model is the involvement of local people, which means there is a greater likelihood of it meeting various local objectives.

Assessing Livelihood Resilience

Resilience is not fixed but changes over time and across contexts. When analyzing resilience, it should be contextually and temporally specific. It is impossible to set a list of indicators to assess livelihood resilience, and so the term "surrogates" is suggested to be used (Nyamwanza 2012).

In the lens of resilience livelihood, CBT is considered in this study as one of the pathways indigenous people choose to cope with livelihood stress. Next, the study examines the characteristics of tourism development at the study sites and the participation of communities in tourism based on the mobilization of their assets. Given the features of the tourism industry, different assets of the sustainable livelihoods framework (The Department for International Development 1999) will be modified to suitably apply in this study: human capital, social capital, tourism resources (natural capital, cultural capital), and other capital (political condition and position). This livelihood capital not only includes endowments (resources, assets) but also refers to entitlement, which means the actor's access to resources (Ifejika Speranza et al. 2014).

Livelihood capital portrays the buffer capacity, which forms the first indicator to assess livelihood resilience. Besides the buffer capital, "self-organization" is identified as another dimension of resilience (Ifejika Speranza et al. 2014). Following the definition of some scholars, we understand "self-organization" as "the internal controllability of a system," which implies a certain level of autonomy, freedom to act, collective action, self-help, self-reliance, independence, power, and control of the own action, all of which can foster identity, trust, and confidence and contribute to empowerment (Holling 2001; Ifejika Speranza et al. 2014). Under the specific context of developing CBT, self-organization highlights the ability of indigenous people to be involved in all aspects (social, economic, and environment) while empowering themselves through tourism activities. The degree of self-organization in this situation is interpreted by the existence of institutions (rules, local norms, and customary laws), the strong network structure and cooperation, and the ability to self-manage tourism activities or the level of participation in tourism (passive involvement, active involvement, full involvement) (Ashley and Roe 1998).

1.2.5 Synthesis Framework

As a result of the literature review, a conceptual framework was developed and modified to meet the specific context in Vietnam's Central Highlands as well as this study's objectives. This framework is used as a thinking tool to outline the key factors that need to be addressed for assessing the vulnerability context of land-use change and livelihood constraint in Vietnam's Central Highlands.

Vulnerability is identified as the main conceptual foundation of this study. It creates a platform for a macro-level examining of external sociopolitical and land-use change. After that the study zooms in on the microlevel – at household scale to research internal sensitivity. Resilience livelihood is the last concept examined, providing insight into one of the ways the community has chosen to cope with the constraints on their livelihood. Three concepts are compiled into an analytical framework, and they are interpreted in more detail using a set of indicators that have been generated based on contextual and temporal adoption (Fig. 1.6).

1.3 Methodology

The research aims to investigate the vulnerability context in the Central Highlands of Vietnam, which is hypothetically denoted by land-use change and socioeconomic transitions over time, especially after the Vietnam War. Moreover, specific internal drivers attributing the indigenes' livelihood pathways as a reaction of pressure contexts will be examined. The study, therefore, calls for a combination of spatial data and survey data, which are obtained through a linear process of the following steps: (1) collecting and analyzing spatial data and determining study sites, (2) collecting secondary data, (3) carrying out key informant interviews and in-depth interviews at households, and (4) computing, aggregating, and analyzing collected data (Fig. 1.7).

1.3.1 Land Use-Land-Cover (LULC) Change and Remote Sensing

The change in land use was evaluated in spatial aspect based on satellite images. In this study, LULC change first gave a backdrop of common transformations in the research area for the last period. It was also an element to decide study sites. Finally, a map of land use at present described various sceneries resulting from human employments of that land.

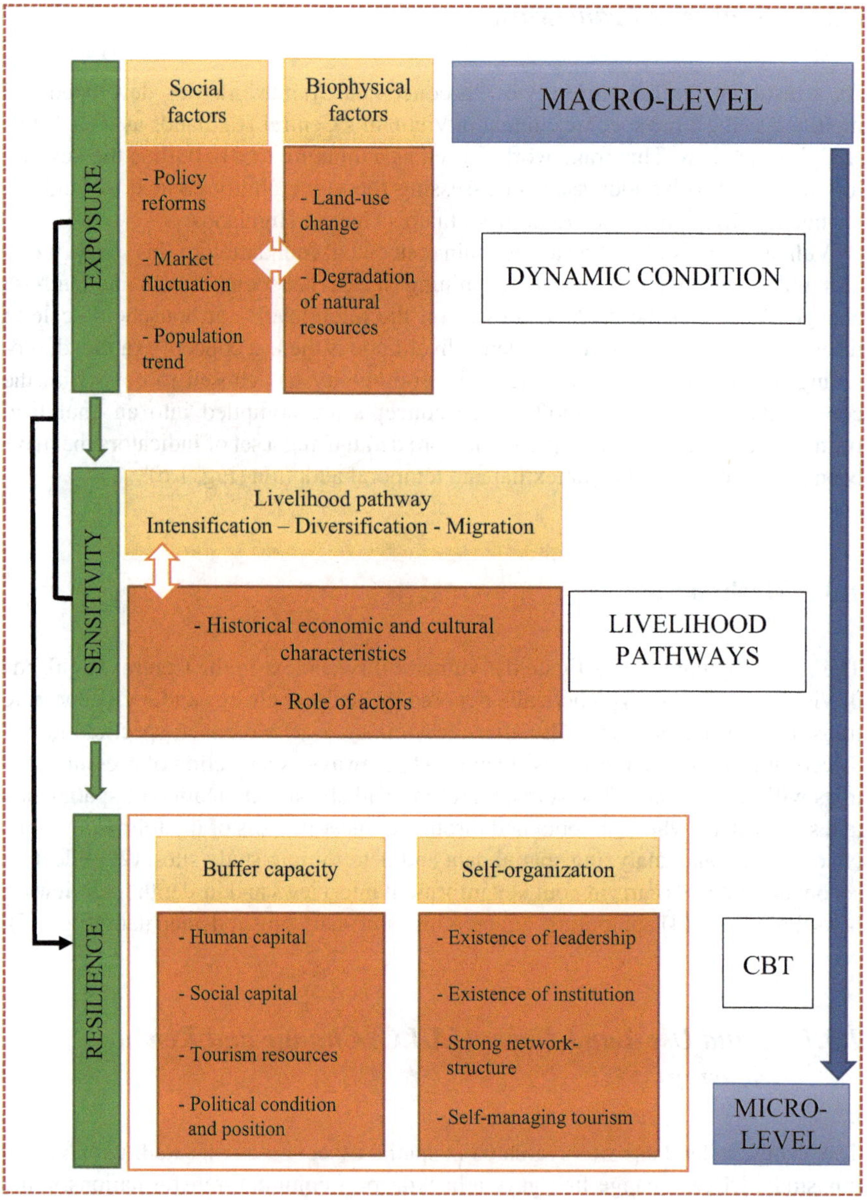

Fig. 1.6 Synthesis conceptual framework (*Source*: Huỳnh Anh Chi Thái)

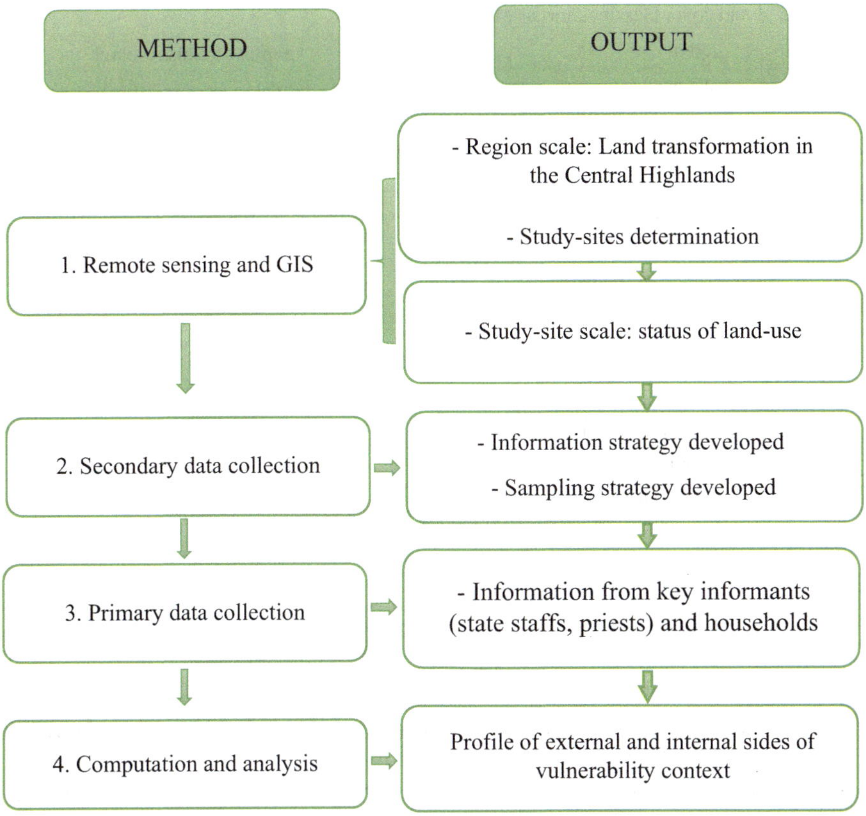

Fig. 1.7 Flowchart illustrating the methodological sequence of research elements (*Source*: Huỳnh Anh Chi Thái)

Land Use-Land Cover (LULC) Change in the Region Level

Maps of LULC of 2000, 2005, and 2010 were used in this study to give insights into socioeconomic changes at the study sites. Actually, these changes started especially after the reunification of the country (in 1975). For this early time, Landsat[3] images are the only one available at 30–120 m spatial resolution. However, many of these old images are full of noise and cloud covered. Moreover, the Central Highlands are a big area of 54.641 km^2. The detection of LULC change of this region, in a period of more than 40 years (from 1975 to present), is a really huge and challenging workload to gain exact results.

[3] Landsat imagery is available since 1972 from six satellites in the Landsat series. These satellites have been a major component of NASA's Earth observation program, with three primary sensors evolving over 30 years: MSS (Multispectral Scanner), TM (Thematic Mapper), and ETM+ (Enhanced Thematic Mapper Plus) (http://glcf.umd.edu/data/landsat/).

Table 1.2 Land cover classification of MCD12Q1 product

Class	IGBP	Adopting to this research
0	Water	Forest
1	Evergreen needleleaf forest	
2	Evergreen broadleaf forest	
3	Deciduous needleleaf forest	
4	Deciduous broadleaf forest	
5	Mixed forest	
6	Closed shrub lands	Shrub lands
7	Open shrub lands	
8	Woody savannas	Savannas
9	Savannas	
10	Grasslands	Grassland/barren
11	Permanent wetlands	Wetlands
12	Croplands	Croplands
13	Urban and built-up	Urban
14	Cropland/natural vegetation mosaic	Croplands
15	Snow and ice	[Not available]
16	Barren or sparsely vegetated	Grassland/barren
254	Unclassified	[Not available]
255	Fill Value	[Not available]

In order to solve these technical problems, the author decided to adopt available data, so-called MODIS Land Cover Type product[4] (MCD12Q1), which supplies global maps of LULC from 2000, into this research. International Geosphere-Biosphere Programme (IGBP) classification of this product was used. However, some classes were merged together to be suitable with the research's aim. Land cover maps after that were imported in the ArcGIS 10.1 software to add the delineation of the Central Highlands (as it was in 2015). The results are maps of LULC change of this region for 2000, 2005, and 2010. These maps were supplemented by statistical data to give an overlook over this huge area (Table 1.2).

Selection of Research Area and Sites

Change in land and socioeconomic status was the common transformation in the Central Highlands, but in different speed among provinces. In order to reveal multifaceted influences of transformations on indigenes and shed light to their dynamic

[4]MCD12Q1 is derived from observations spanning a year's input of Terra and Aqua MODIS data at 500 m spatial resolution. Many components of the classification algorithm are applied: training site is revised, land surface temperature is included, and ancillary datasets are used in post-processing (Friedl et al. 2010). The interpreting process, after that, was conducted by a supervised classification. The primary land cover scheme identifies 17 land cover classes defined by the International Geosphere Biosphere Programme (IGBP), which includes 11 natural vegetation classes, 3 developed and mosaicked land classes, and 3 nonvegetated land classes. For ease of use, a number of other classification schemes are also provided by different communities (Friedl and Sulla-Menashe 2013).

reactions to different levels of land constraints, study sites were purposely picked at disparate rates of LULC change (Table 1.3).

The statistical data combined with land-use map illustrates the changes in LULC among provinces in the Central Highlands. Out of them, Dak Lak is remarkable with the drastic expansion of agricultural land and the severe loss of forest. Meanwhile, the transformation in Kontum and Lam Dong is moderate. These provinces convey disparate landscapes and will be primarily chosen as research areas.

Moreover, given the research aim to learn about indigenes' livelihood pathway, the existence of various livelihood options apart from agriculture, in particular the availability of tourism activities, appears as the second criterion of study site selection. Based on primary information of these uplands and economic activity characteristics of the indigenes (Hickey 1967; McElwee 2008b; United States Agency for International Development 2008), it indicates that the remoter the area in which the indigenes lived, the more they depend on the land and forests. Multi-choice of livelihood could be met at the not-so-much isolated areas.

For all above, Kontum City (Kontum Province), Lak District (Dak Lak Province), and Lac Duong District (Lam Dong Province) met these criteria and were set as the study sites for the subsequent steps of this research (Fig. 1.8).

Various Land-Use Sceneries at Study Sites

At study site scale, Landsat satellite images were used to detect the sceneries of contemporary land use. Landsat 8 OLI/TIRS (Operational Land Imager/Thermal Infrared Sensor) was downloaded from the US Geological Survey archives (http://earthexplorer.usgs.gov/). The selected images were at the path 125, row 50-51-52, respectively, with areas of Kontum City, Lak District, and Lac Duong District. All images were taken at the same period in 2015 (dated from 04 March to 22 March), which were the latest images at the time the data analysis was carried out. The data was relatively clear and cloud free.

The image interpretation was operated by ENVI 4.7 program. The technique to derive land cover information was run under supervised classification.[5] The author applied the maximum likelihood method to the five corrected bands of each image. Image interpretation was based on training sites, which had been collected at study sites during a field trip from October 2014 to February 2015. Based on the objectives of this research and local conditions, seven classes were identified to be detected from satellite images. They are (1) paddy/or flower,[6] (2) annual land, (3) perennial land, (4) forest, (5) settlement, (6) water, (7) bare land, and (0) unclassified.

[5] Supervised classification is based on the spectral characteristics of identified sample areas or training fields (Daniel Müller 2003).

[6] Due to specific crops among different study sites, paddy class will be detected in Kontum City and Lak District, and flower class will be detected in Lac Duong, instead.

Table 1.3 Change in land use in the Central Highlands during period 1995–2007

		1995	1998	1999	2000	2005	2006	2007	Change 1995–2007
Kontum	Agriculture land	51,300	69,000	82,600	92,300	128,400	129,100	135,400	**84,100**
	Forestry land	634,300	628,400	612,700	606,700	662,900	662,500	660,300	**26,000**
Gialai	Agriculture land	244,100	277,300	315,600	375,500	499,500	499,700	499,500	**255,400**
	Forestry land	846,800	858,200	813,600	750,800	799,800	796,900	799,800	**−47,000**
Dak Lak[a]	Agriculture land	328,900	392,300	446,300	524,900	688,300	690,900	692,200	**363,300**
	Forestry land	1,209,800	1,154,900	1,100,100	1,017,900	988,700	986,500	983,600	**−226,200**
Lam Dong	Agriculture land	184,200	204,100	222,300	240,900	277,500	277,100	277,100	**92,900**
	Forestry land	552,000	598,200	614,700	617,800	622,300	622,400	622,400	**70,400**

Source: (General statistical office 2010)

[a]Dak Lak Province was divided into Dak Lak and Dak Nong in 2004. To learn about the land transformation among provinces during 1995–2007, we merge the data of Dak Nong from 2005 into Dak Lak

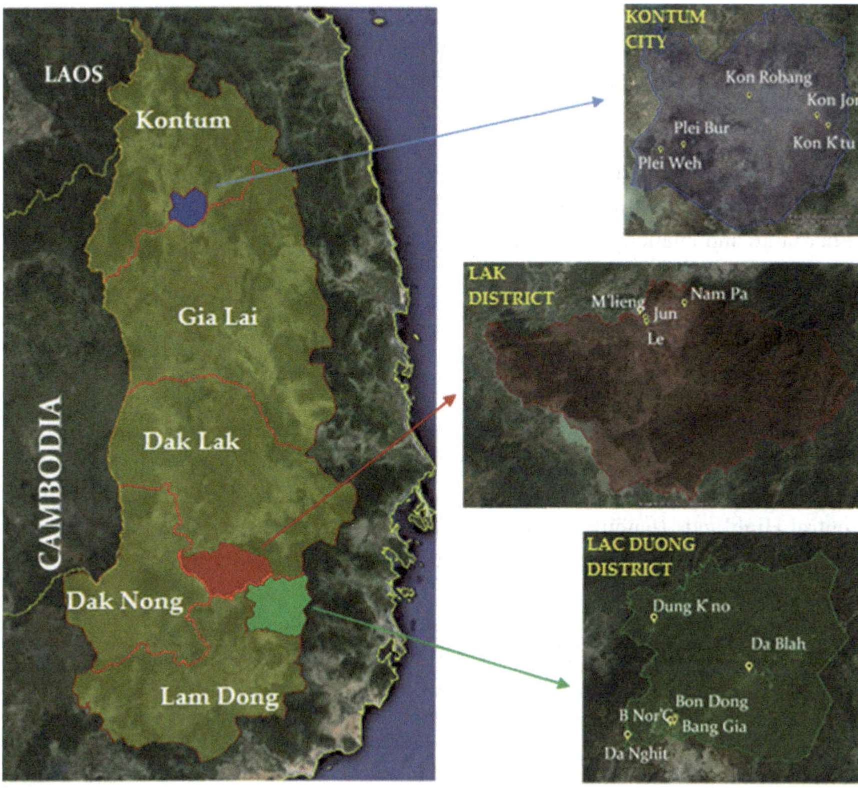

Fig. 1.8 Study sites in Vietnam's Central Highlands

1.3.2 Secondary Data

Secondary data included statistics, annual reports, working paper, scientific reports, and maps (administration maps, land-use maps) which were collected from government agencies and libraries in Kontum, Dak Lak, and Lam Dong (at both provincial and district levels).

Collecting secondary data was repeated as the first step of every study sites access. Normally, we contacted the People' Committee offices first to get the general official information about study sites, which was issued as annual reports. These reports have been collected since 2010. It was difficult to access the older data as annual reports were usually not archived systematically for long term, in particularly at sub-administration like district. Exceptionally, statistical books were issued in Lak District that offered systematic data from 2005 onward.

Besides, more detailed data (such as data about natural conditions, livelihood systems, and ethnic groups) were collected from specific offices, including the Department of Resources and Environment, the Department of Economy, and the Department of Ethnic Minorities. Additionally, reference books and literature review articles were collected at the local library supplying more knowledge about historical records or special facts and figures.

The same statistics data were collected from different sources offering a chance to compare, generalize, and aggregate data. Later, the data was cross-checked again with the interview data. This approach is called methodological triangulation, aiming at corroborating and augmenting evidences from other sources (Decrop 1999). Secondary data was important to get a holistic context of study sites, which was rather large and challenging to access on the field trip around.

1.3.3 Case Study Methodology

Given the aims to study about internal factors driving livelihood pathways of indigenes, the concerned information may not be acquired by quantitative survey. Furthermore, due to demographical characteristics of indigenous community in the Central Highlands (language barrier, high level of illiteracy), some static research instruments unlikely gain adequately crucial data. Therefore, the case study is a reasonable methodology to be applied, and the in-depth focus on the cases also covers a broad range of complex conditions that makes the case study go beyond the study of isolated variables (Yin 2013, p.4).

Research Setting

The field research was set up and conducted through different steps

Interview Specialists for First Patterns and Concepts

The Central Highlands is said to be a tough research area.[7] In order to set up field research, it is necessary to establish a network of information, which started with some contact persons who are experienced in doing research in these uplands and could provide "tips" for further research.

For the patterns and some first information about the research area, the author did interviews with university professors, scientists who did previous research on the Central Highlands, state staffs, and also priests who service in this area and have had close contacts with the indigenous communities for a long time.

[7]Vietnam's Central Highlands is a tough study area, as it is a huge mountainous area where it is historically viewed as a remote, isolated area that is difficult to access. Moreover, in recent decades, Central Highlands has come out to be a sensitive area due to some conflicts and demonstrations relating to land rights and freedom of religion in 2001 and 2004 (Human Rights Watch 2002; UNHCR 2006). Therefore, research about Central Highlands, in whatever topics, is regarded as socially or politically sensitive.

Accessibility to the Field

With the background information, the research was started from the top of state hierarchy and then worked down to village level. As usual, it is necessary to legitimate the researcher's presence in the field (Huynh 2015), thus the author's identity should be clear before setting up the field work. Although the study is not on a politically or socially sensitive topic, due to the cautiousness about foreign factors of Vietnamese authority, the author could not appear on the field as a PhD student coming back from a Western university. Instead, taking advantage of being a staff at a state university in Vietnam, the author was first authorized to be a researcher under Ho Chi Minh City University of Culture. The introduction letters from this university send her to different state offices in Kontum, Dak Lak, and Lam Dong. Then she could move down from province to district and commune level.[8]

Case Study Definition

In this study, the household was supposed as a unit to be examined. Although household members might have diverse livelihood decisions, these decisions supported and influenced each other. Consequently, livelihood pathways were viewed as the result of a decision-making process which created a portfolio of activities.

Although indigenous groups live in matriarchal communities, in most cases, husbands were interviewed. In this modern society, men became the main laborers and had more power to make an economical decision. However, in some cases, both husband and wife were interviewed at the same time to provide the most thorough recall on answers. Given the initial ambitious research aims, the historical stories since 1975 would have liked to be revived. However, a course of 40 years (from 1975 to present) is rather long. Local people who witnessed the whole process of transformations were the old people. They neither spoke Vietnamese nor were lucid enough to remember and narrate these events. Meanwhile, the young people did not care or were not fully aware of the affairs happening before they were born. Thus, while consequences of these events were still remaining so far, the rooted reasons were not clearly indicated and considered by local people.

[8] At every study site, the author had to firstly present herself at the highest government management offices (the People's Committee) and obtained the contacts with local cadres (in communes and villages). A village in the Central Highlands nowadays is a state administrative unit though it still appears as a traditional social unit controlled by its header. The mixture in the management of its society generated more barriers and confusions to access. The experiences from the specialists introduced that besides asking for the permission from authorities, the researchers should get contact with villages' headers on the field as the first step to "penetrate" study sites.

Characteristics of the Selected Cases of Communes

Study sites had been purposely chosen in previous steps including Kontum City, Lak District, and Lac Duong District. In order to set up in-depth interviews, the study sites were progressively narrowed down to certain villages. The criteria for choosing study sites should be also met in microscale (villages and even households) as well. Therefore, chosen villages are usually located at the central of district/ city or not too far from the center where economic activities were various and tourism potential was given (Table 1.4).

Combination of Observations and Interviews

Accompanied by an assistant, the author strolled through the villages every day. The observations were focused mostly on economic activities of indigenes. Apart from diverse human activities, the natural landscapes relating to villagers' lives were also observed to get an aggregated analysis of the human-nature relationship. During these observations, GPS was used to record the location of study sites and, even more importantly, to collect the ROI (region of interest or types of land use) which would be used to analyze satellite images after that.

The main issues covered in the interviews ranged from land use and its influences to indigenes' livelihood transformation. The number of sampling was rather balanced between the study sites. At the first site – Lac Duong District – the interviews were conducted until reaching data saturation, as more case studies did not bring any additional insights into the research questions (Mack et al. 2005). When moving to the other study sites (Lak District and Kontum City), samples were decided to collect correspondingly to the sample size of the first site. Additionally, the snowball sampling method was used. Based on the community networks, par-

Table 1.4 Characteristics of the study sites

		Level of livelihood dynamic[a]		Villages with existence of tourism[b]	
District/ city	Level of land transition	Existence of mono-annual crops	Level of service sector activities	Yes	No
Kontum	Low	Yes	Low	Kon K'tu, Kon Jo Ri	Kon Ro Bang, Plei Weh
Lak	High	Yes	Moderate	Le, Jun	M'Lieng, Nam Pa
Lac Duong	Moderate	No	High	Bang Gia, Bon Dong	Da Blah

[a]According to the developed stages of rural economic activities in the Central Highlands, mono-annual crops are placed at the lowest step, whereas service sector (trade, tourism activities) is set at the highest stage

[b]All of these villages have been known by tourists. However, the existence of tourism is identified by the presence of tourism station on site or at least the tourism participation of local people

ticipants could recommend their co-villagers who likely participated in the study. This helped to save the time of walking around to find suitable samplings, and more importantly, the trust built in some first contacts could be spread further.

The interviews began by some icebreaking components, as the author told local people about her background and shared her concerns about this area which were sometimes beyond the research questions. In this way, these stories softened the contacts, encouraging participants to open up. Identities of informants were treated as private property and would not be directly mentioned in this work. The interview sites were mostly at home; they sometimes included fields or tourism sites, where the author could simultaneously observe local economic activities. The field trip was conducted from October 2014 to February 2015. During this time (especially in October, November) indigenous farmers were rather busy with their harvests. In some days, villages were empty, as villagers all gathered on their farms which were usually far away from their settlements. This situation encouraged the author to follow them to the fields; henceforth, data was collected on working days, during lunch break, or even during working time. The length of each interview might last a course of several hours. A standardized interview was 90 min.

All the responses were noted on paper. The research assistant tried to keep the notes as original as possible. She wrote down almost exactly every word the participants responded. Moreover, we used tape recording to record the answers. The records after that were treated and nonessential parts such as jokes or gossip were cut off. Then the treated records were all transcribed. These documents (the notes and the transcriptions) were cross-checked before the data was coded. At the end, all collected data included jotted notes, direct observation notes, transcriptions from recording, and a fieldwork diary. This qualitative data was gathered from 57 interviews, and of them, 49 interviews were held with households and 8 with specialists. Due to a huge load of transcriptions, it was impossible to translate all of them from Vietnamese to English. The author decided to keep them in the original, and a certain amount of data which was important would be translated into English later as they were quoted into the research.

Analyzed Data and MAXQDA Program

After transcribing and refining, the data was inserted into MAXQDA where the next steps of coding and interpreting data would take place (Table 1.5).

Based on the vulnerability concept and its dimensions, the interview transcripts have been conceptualized through interconnected codes. Data from three study sites were coded simultaneously and used within the same coding system. While the exposure dimension was examined from region level as a common context, the remaining codes allowed a comparison across three study sites when we zoomed into the household level.

Table 1.5 Dimension and subdimension for coding qualitative data

	Dimension	Subdimension
1	Vulnerable context (exposure)	Perception of land-use change
2		Influence of vulnerable context
3		Relationship with land, farm
4	Sensitivity in livelihood pathway	Common characteristics
5		Actor role
6	Livelihood resilience	Human capital
7		Social capital
8		Self-organization

References

Adger WN (2006) Vulnerability. Glob Environ Chang 16:268–281. https://doi.org/10.1016/j.gloenvcha.2006.02.006

Ashley C, Roe D (1998) Enhancing community involvement in wildlife tourism: issues and challenges. International Institute for Environment and Development, London

Birkmann J (2006) Measuring vulnerability to promote disaster-resilient societies: conceptual frameworks and definitions. In: Birkmann J (ed) Measuring vulnerability to natural hazards: towards disaster resilient societies. United Nations University Press, Tokyo/New York/Paris, pp 9–54

Birkmann J, Cardona OD, Carreño ML, Barbat AH, Pelling M, Schneiderbauer S et al (2013) Framing vulnerability, risk and societal responses: the MOVE framework. Nat Hazards 67(2):193–211. https://doi.org/10.1007/s11069-013-0558-5

Bohle H-G (2001) Vulnerability and criticality. IHDP Updqate 2/2001. Newsletter of the International Human Dimensions Programme on Global Environmental Change

Bourdieu P (1977) Outline of a theory of practice. In: Gellner E, Goody J, Gudeman S, Herzfeld M, Parry J (eds) Cambridge studies in social and cultural anthropology. Cambrige University Press, Cambridge

Brooks N (2003) Vulnerability, risk and adaptation: a conceptual framework (Tyndall Centre Working Paper No. 38). Tyndall Centre Working Paper

Chambers R (2006) Vulnerability, coping and policy (editorial introduction). IDS Bull 37(4):33–40

Coleman JS (1988) Social capital in the creation of human capital. Am J Sociol 94:95–120

Cutter SL (1996) Vulnerability to environmental hazard. Prog Hum Geogr 20(4):529–539

Dang TH, Espaldon MVO (2001) Balancing economic and environmental concerns in the pplands of Vietnam: a continuing challenge. SANREM CRSP Research Scientific Synthesis Conference, Athens

Dang TH, Shively G (2005) Coffee vs. cacao: a case study from the Vietnamese central highlands. J Nat Res Life Sci Edu 34:107–111

De Chazal J (2010) Examining resilience and vulnerability as concepts conditional upon human values: a review. Canberra, Australia.

Decrop A (1999) Triangulation in qualitative tourism research. Tour Manag 20(1):157–161. https://doi.org/10.1016/S0261-5177(98)00102-2

Department for International Development (1999) Sustainable livelihoods guidance sheets. Department for International Development, London

Devereux S (2001) Sen's entitlement approach: critiques and counter-critiques. Oxf Dev Stud 29(3):245–263. https://doi.org/10.1080/13600810120088859

Ellis F (1999) Rural livelihood diversity in developing countries : evidence and policy implications. ODI Nat Res Perspect 40:1–10

Espaldon DMVO, Phuoc PHD, Thuy NN, Du L Van, Hung PT, Thong LQ, … Ha DT (2004) Challenges on sustainable agriculture and natural resource management in Vietnam uplands: a case study. SEAMEO SEARCA

Friedl M, Sulla-Menashe D (2013) User guide for the MODIS land cover type product (MCD12Q1). https://doi.org/10.1017/CBO9781107415324.004

Friedl MA, Sulla-Menashe D, Tan B, Schneider A, Ramankutty N, Sibley A, Huang X (2010) MODIS collection 5 global land cover: algorithm refinements and characterization of new datasets. Remote Sens Environ 114:168–182. https://doi.org/10.1016/j.rse.2009.08.016

Fünfgeld H (2007) Fishing in muddy waters: socio-environemental relations under the impact of violence in eastern Sri Lanka. Verlag für Entwichlungspolitik, Saarbrücken

Füssel H-M (2007) Vulnerability: a generally applicable conceptual framework for climate change research. Glob Environ Chang 17:155–167. https://doi.org/10.1016/j.gloenvcha.2006.05.002

General Statistical Office (2010) Socio-economic statistical data of 63 Provinces and cities. Statistical Publishing House, Hanoi

General Statistical Office (2015) Statistical yearbook of Vietnam. Statistical Publishing House

Giddens A (1984) The constitution of society: outline of the theory of structuration. University of California Press, Berkeley/Los Angeles

Hamzah A, Khalifah Z (2009) Handbook on community based tourism: "How to develop and sustain CBT". Asia-Pacific Economic Cooperation, Kuala Lumpur

Hardy A, Turner S (2000) Editorial: migration, markets and social change in the highlands of Vietnam. Asia Pacific Viewpoint 41(1):1–6

Hickey GC (1967) The highland people of South Vietnam: Social and economic development. The Rand Corporation, Washington, DC

Holling CS (2001) Understanding the complexity of economic, ecological, and social systems. Ecosystems 4:390–405. https://doi.org/10.1007/s10021-001-0101-5

Human Rights Watch (2002) Repression of Montagnards: conflicts over land and religion in Vietnam's Central Highlands. Washington/London/Brussels, New York

Hussein K, Nelson J (1999) Sustainable livelihoods and livelihoods diversification (IDS Working Paper 69 No. 69)

Huynh TPL (2015) State – society interaction in Vietnam: the everyday dialogue of local irrigation management in the Mekong Delta. Bonn University, Bonn

Ifejika Speranza C, Wiesmann U, Rist S (2014) An indicator framework for assessing livelihood resilience in the context of social–ecological dynamics. Glob Environ Chang 28:109–119. https://doi.org/10.1016/j.gloenvcha.2014.06.005

ITTO,&IUCN(2005)Restoringforestlandscapes:anintroductiontotheartandscienceofforestlandscape restoration. ITTO technical series no 23. https://doi.org/10.1890/1540-9295(2005)003[0243:ROFPCI]2.0.CO;2

Klein RJT, Nicholls RJ (1999) Assessment of coastal vulnerability to climate change. Ambio 28(2):182–187. https://doi.org/10.2307/4314873

Leach M, Mearns R, Scoones I (1999) Environmental entitlements: dynamics and institutions in community-based natural resource management. World Dev 27(2):225–247

Leisz SJ, Fox J, Yasuyuki K, Masayuki Y, Rambo A, T. (2009) Land use changes in the uplands of Southeast Asia: proximate and distant causes. Southeast Asian Stud 47(3):237–243

Mack N, Woodsong C, McQueen KM, Guest G, Namey E (2005) Qualitative research methods: a data collector's field guide, vol 36. Family Health International, Triangle Park. https://doi.org/10.1108/eb020723

McElwee P (2001) Coffee, Christianity and conflict: an update on the central highlands of Vietnam. Mekong Update Dialogue 4(3):8

McElwee P (2008a) Becoming Socialist or becoming Kinh. In: Duncan CR (ed) Civilizing the margins: Southeast Asian government policies for the development of minorities. NUS Press, Singapore

McElwee P (2008b) "Blood relatives" or uneasy neighbors? Kinh migrant and ethnic minority interactions in the Trường Sơn Mountains. J Vietnam Stud 3(3):81–116. https://doi.org/10.1525/vs.2008.3.3.81

Moser CON (1998) Reassessing urban poverty reduction strategies: the asset vulnerability framework. World Dev 26(1):1–19

Müller D (2003) Land-use change in the Central Highlands of Vietnam a spatial econometric model combining satellite imagery and village survey data. Georg-August-Universität Göttingen, Göttingen

Müller D, Zeller M (2002) Land use dynamics in the central highlands of Vietnam: a spatial model combining village survey data with satellite imagery interpretation. Agric Econ 27:21

Niehof A (2004) The significance of diversification for rural livelihood systems. Food Policy 29:321–338. https://doi.org/10.1016/j.foodpol.2004.07.009

Nyamwanza AM (2012) Livelihood resilience and adaptive capacity: a critical conceptual review. Jàmbá J Disaster Risk Stud 4(1):1–6. https://doi.org/10.4102/jamba.v4i1.55

Parker J (2010) Structuration theories. Historical developments and theoretical approaches in sociology, vol 2. https://doi.org/10.4324/9780203519394.ch4

Parry M, Canziani O, Palutikof J, Van der Linden P, Hanson C (2007) Climate change 2007: impacts, adaptation and vulnerability (Intergovernmental panel on climate change). Cambridge University Press, Cambridge

Pelling M (2010) Adaptation to climate change: from resilience to transformation. Routledge, London/New York

Prowse M (2003) Towards a clearer understanding of "vulnerability " in relation to chronic poverty. Chronic Poverty Research Centre No. 24

Rietbergen-McCracken J, Maginnis S, Sarre A (2007) The forest landscape restoration handbook. In: Sayer JA (ed) The Earthscan forestry library. Earthscan, London

Sen A (1981) Ingredients of famine analysis: availability and entitlements. Q J Econ 96(3):433–464

Smith T, Lynam T (2010) Towards enhancing adaptive capacity for climate change response in South East Queensland. Australas J Disast Trauma Stud 1

Sunderlin WD, Ba HT (2005) Poverty alleviation and forests in Vietnam. Center for International Forest Research (CIFOR), Bogor Barat

Tahmasebi A (2012) Pastoralism under pressure: vulnerability of pastoral nomads to multiple socio-political and climate stresses – The Shahsevan of Northwest Iran. Bonn University, Bonn

The World Bank (2009) Country social analysis: ethnicity and development in Vietnam, vol 1 and 2. Social Development Unit, East Asia and Pacific Region, Washington, DC

Thywissen K (2004) Comparative glossary for core terms of disaster reduction. Institute for Environment and Human Security, United Nations University, Bonn

Tran TPH (2012) Resilience and livelihood dynamics of shrimp farmers and fishers in the Mekong Delta, Vietnam. Wageningen University, Wageningen

Turner BL, Kasperson RE, Matson PA, McCarthy JJ, Corell RW, Christensen L et al (2003) A framework for vulnerability analysis in sustainability science. Proc Natl Acad Sci U S A 100(14). https://doi.org/10.1073/pnas.1231335100

UNHCR (2006) Vietnam: situation of indigenous minority groups in the Central Highlands

United States Agency for International Development. (2008) Vietnam Central Highlands needs assessment. USAID

Van Dillen S (2004) Different choices: assessing vulnerability in a south Indian village. Verlag für Entwichlungspolitik, Saarbrücken

Watts M, Bohle H (1993) The space of vulnerability: the causal structure of hunger and famine. Prog Hum Geogr 17(1):43–67

Wells-Dang A (2012) Ethnic minority development in Vietnam: what leads to success? Background paper prepared for the 2012 Poverty Assessment, Hanoi, Apri, Hanoi

Wiesmann U, Ott C, Speranza CI, Kiteme BP, Müller-Böker U, Messerli P, Zinsstag J (2011) A Human actor model as a conceptual orientation in interdisciplinary research for sustainable development. In: Wiesmann U, Hurni H, with an international group of co-editors (eds) Research for sustainable development: foundations, experiences, and perspectives of the Swiss National Centre of Competence in Research (NCCR) North-South. University of Bern, Bern, pp 231–256

Wisner B, Blaikie P, Cannon T, Davis I (2003) At risk second edition: natural Hazards, people's vulnerability and disasters. Routledge, London/New York

Wolf S, Hinkel J, Hallier M, Bisaro A, Lincke D, Ionescu C, Klein RJT (2013) Clarifying vulnerability definitions and assessments using formalisation. Int J Clim Change Strategies Manage 5(1):54–70. https://doi.org/10.1108/17568691311299363

Yin R (2013) Case study research: design and methods. Sage, Thousand Oaks

Chapter 2
Exposure Context: Socioeconomic Transformations and Land-Use Change

Abstract The vulnerability context of indigenous people is driven by natural and social factors. Of these factors, land-use change is considered one of the major driving forces causing the constrained status of indigenous people. Other trends, including massive migration and market change, contribute to indigenous people's vulnerability. This chapter deals with an integrated examination of this multifacet context in the Central Highlands. The context is described based on various sources of historical digital data (satellite images and maps), secondary data, and historical information from interviewees. The results present a common picture of land use, which depicts the consequence of historical factors, and various contemporary scenes, which reflect the dynamic livelihoods at the study sites. Besides, the perception of indigenous people about their vulnerability context is revealed, which mainly relates to land-use change and is recognized as driving force motivating them to make change in their livelihoods.

Keywords Land-use change · Migration · Coffee market change · Vulnerability context · External factors

2.1 Political and Economic Conditions

This study pays primary attention to the policies implemented in the Central Highlands after the Vietnam War, which dramatically changed the socioeconomic status of indigenous people. Some policy plans, however, are rooted in the previous period. One of the most important and ambitious plans implemented at that time involved land tenure and population resettlement.

H.A.C. Thái, *Livelihood Pathways of Indigenous People in Vietnam's Central Highlands*, Advances in Asian Human-Environmental Research,
https://doi.org/10.1007/978-3-319-71171-3_2

2.1.1 Land Tenure

After Vietnam's reunification, the Central Highlands was seen as a frontier for the economic development (Espaldon et al. 2004). In order to urgently address the mass starvation and achieve a balance in the food production between the north and the south of the country, a series of policies was implemented to restart agriculture on abandoned and war-torn lands (UNHCR 2006). This, in turn, led to massive transformations in this upland area.

During the first step of this transformation period, which involved an ambitious plan to rapidly implement socialism in the former South Vietnam, economic and political measures enforced in the north since the 1950s were exported to the south (Michaud 2009). Individual land ownership was then banned, whereas collectivization[1] and migrations were applied (Cheesman and Bennett 2005). The new economic zones were officially commended at the Fourth Party Congress in 1976 and incorporated into the Second Five Year Plan (1976–1980) (UNHCR 2002). Consequently, new state-owned farms (SFs – *nông trường)* and state forest enterprises (SFEs – *lâm trường)* were established. By 1988, there were 83 SFEs (accounting for three million hectares of forest land – 70% of the area) and 79 SFs in the Central Highlands (UNHCR 2006). Simultaneously, a long-term population de-urbanization and re-distribution were initiated from the north and urban centers in the south to the Central Highlands under a state-sponsored transmigration program (Dang et al. 2001; Cheesman and Bennett 2005; UNHCR 2006). The resettlement of migrants was accompanied by the sedentarization of the indigenous population. As the result, the Central Highlands witnessed a remarkable socioeconomic and environment transformation. The population rose sharply, and the population composition and social structure changed dramatically. Large forest areas were replaced by agricultural areas. During this period, none of the indigenous people's traditional land rights or traditional managements of natural resource systems were recognized, because they were thought to be rooted in archaic technology. Instead, subsidiary state-run agricultural farms and forest enterprises were employed (Espaldon et al. 2004). Collectivization in the end failed to produce the expected outcome and resulted in economic stagnation, hyperinflation, and food shortages (Cheesman and Bennett 2005). This prompted new reforms, which included the dismantlement of agricultural collectives.

In the early 1980s, Vietnam shifted from a centrally planned to a state-regulated market-oriented economy, and a series of institutional and policy reforms was adopted (Espaldon et al. 2004). Notable programs were the reforms in the agricultural sector supported by the institutional changes which comprised the implementation of the contract systems embodied in Directive No.100 in 1981. According to

[1]Agricultural collectivization along with central planning and heavy industry was recognized as a "Soviet model" (Hill 2014) and was started in North Vietnam (or the Democratic Republic of Vietnam) in 1958.

this system, lands initially allocated by the state to cooperatives were reallocated to individual households for family farming.

In 1986 at the Sixth Party Congress, the state officially began the economic reforms known as *Đổi Mới*, which proclaimed the household to be the main economic unit in agriculture (UNHCR 2002). The shift from a cooperative to household management system continued with the removal of collective ownership of agricultural lands and production capital through the Central Committee Resolution 10 "Renovation of Management of the Agricultural Sector" in 1988 (Cheesman and Bennett 2005). The significant contents of this de-collectivization policies are *(1) recognition of the farm household as the main unit of agricultural production, (2) liberalization of farm decision-making with respect to the purchase of inputs and sale of outputs and market system was developed, and (3) quasi-private land rights were established by providing households with long-term (15 years) leasehold entitlements* (Dang et al. 2001).

The final policy reform was the 1993 Land Law which stated five rights in agricultural land: the right to be leased, transferred, exchanged, inherited, and used as collateral (Espaldon et al. 2004). By extending the lease period to 20 years for annual croplands and 50 years for perennial and other long-term crops, including forestry, this law consolidated land tenure security.

The policies and in particularly the *Đổi Mới* process were seen as a postulate and necessary for Vietnam to grow economically; however, these policies had their two sides. The land reforms brought a significant transformation to households, which became the basic unit of agricultural production. It was no longer the leaders of collective or state farms but the individual who decided to farm based on market signals. This raised the role of the market, making it the primary means for determining prices of farming inputs and outputs (Espaldon et al. 2004). Mono- and self-sufficient crops were gradually replaced by multi- and commercialized productions. This trend happened throughout Vietnam and was well reflected in the Central Highlands. With a focus on commodity-crop productions, particularly coffee and rubber, and an encouragement of intensive cultivation, the Central Highlands became the biggest coffee plantation in Vietnam and made the country the second largest coffee producer in the world.

Right from the beginning, the Vietnamese government believed that these policies aimed to help minorities develop and that they would overcome the waste of abandoned lands and bring balance to the reunified area. Simultaneously, the nation became united under a socialist regime, and they emphasized the need for development and incorporation of the highlanders into socialism, rather than emphasizing protection of the local customs and respect for local knowledge (McElwee 2008). These programs, which gained success in the northern lowlands, were stereotypically applied in these central uplands without any experiments and without any considerations for local traditions. The policy reforms and the large numbers of migrants who moved to the Central Highlands made the indigenous people's traditional land management systems nearly obsolete. The traditional land ownership concept of the indigenous people was regarded as backward, primitive, and harmful.

The processes of collectivization and then de-collectivization in the Central Highlands were much slower and more complicated than in other parts of Vietnam (McElwee 2008). In the lowland areas, when agricultural cooperatives were dissolved, farmers received back the land that had been taken from them. However, the case in the Central Highlands was different. For centuries, the ethnic groups in the Central Highlands practiced rotational shifting cultivation (which suited the characteristic of the topography and soil in these highlands), and lands were a village's asset and were managed by clan elders. Also, highlanders preserved and protected forests for both subsistence and spiritual purposes. They collectively controlled land in their village territory according to customary laws and had the right to use and inherit land, but not transfer or sell it to outsiders (UNHCR 2006). The private property and commodity rights that policy reforms brought to the land concept were strange notions and were clearly not compatible with the indigenous people's communal land tenure concept.

The main problem appeared more clearly during the de-collectivization process than the collectivization process, as the indigenous people were not allocated the same land they had brought into the agricultural collective (UNHCR 2002). This is because only permanent agricultural land was recognized. Moreover, it was difficult to transform a system of communal land tenure into private land ownership. Another problem in the Central Highlands was that the SFs persistently occupied fertile land for profitable cash crops and leased out portion of land to local households to manage as sharecroppers, which lengthened the period of de-collectivization ((Salemink 2003a), – draft not for quotation). Meanwhile, most of the forest lands remained in the hands of SFEs and were classified as national parks or nature reserves (ibid). According to governmental program No.327_CT issued in 1992, forest lands were also supposed to be allocated to individual households. However, this process was even slower than the allocation of agricultural land. Program No.202_TTg, issued in 1994, followed next which allocated forest lands to the minorities for protection. Households in this program were paid a limited fee of around USD10 per ha per year (Le and Bang 2013) and were not allowed to exploit the land. The indigenous people on the whole no longer benefitted much from the forest and, in turn, were not interested in these programs.

2.1.2 Market Change

With the series of policy reforms discussed above, Vietnam transformed from a centrally planned economy toward a state-regulated, market-oriented economy. Consequently, households freely decided what crops to produce based on the market. The market integration added another significant factor to the development of this rural area and brought dramatic change to the lives of the indigenous farmers.

During this process, the agricultural system underwent a profound shift from mono-crop, self-sufficient production to diversified crops and more commercialized production. Commercial cash crops such as rubber, tea, and mulberries were promoted. Especially given the high motivation by the high market prices, coffee production rapidly expanded in the Central Highlands thanks to the fertile basaltic soil and favorable climate (Espaldon et al. 2004). The fluctuation of coffee prices that followed brought about remarkable socioeconomic and land-use transformations in this upland area.

The Situation of Coffee Market Change

Coffee was first planted in Vietnam in 1857 and gained a modest area of around 19,000 ha in 1975 (Tran 2007). In the early and mid-1990s, the price of coffee rose at an annual average rate of 30 percent (UNHCR 2006) due to the crop loss and damage in Brazil[2] by severe frosts (Cheesman and Bennett 2005). In response to the appeal of coffee market, Vietnam achieved a significant growth in producing coffee. In the course of a decade, Vietnam increased its coffee production by nearly tenfold (from 92,000 ton in 1990 to 802,000 ton in 2000) (International Coffee Organization 2015) and became a significant player in the world trade for Robusta coffee (Fig. 2.1).

By the end of the 1990s, when Vietnam entered into the world coffee market, coffee provided as much as 10% of Vietnam's annual export earnings (Dang and Shively 2008). This country has created a new balance in the arena of coffee-producing country and been a permanent part of the market (Coffee Association

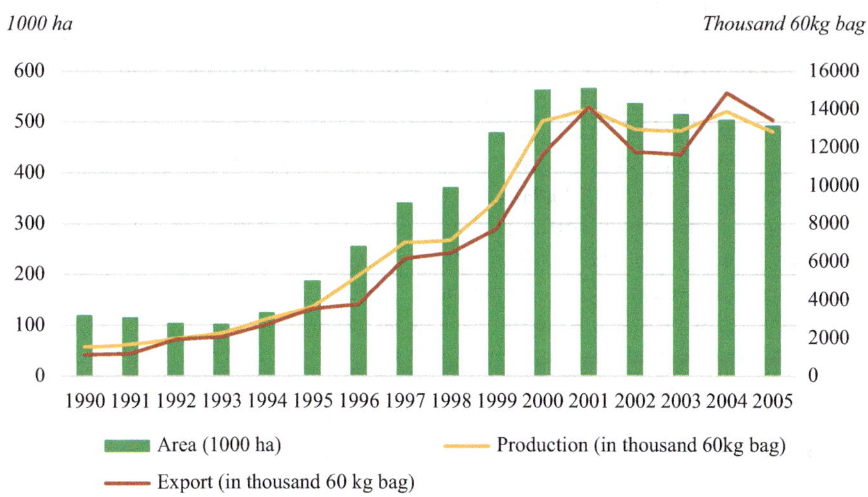

Fig. 2.1 Historical coffee production in Vietnam (1990–2005) (*Source:* Tran 2007; General Statistical Office 2010; International Coffee Organization 2015, drawn by author)

[2] Brazil is a traditional coffee supplier, the top coffee producer, and also one of the main consuming countries in the world (Ponte 2002).

Table 2.1 Exports of all forms of coffee by top five exporting countries (1990–2000) (thousands of 60 kg bags)

	Type of coffee[a]	1990	1995	1997	1999	2000	Share of world production (2000)
Total		80,675	67,872	80,414	86,145	89,559	
Brazil	A/R	16,936	14,468	16,801	23,149	18,016	20.1
Vietnam	**R/A**	**1,145**	**3,546**	**6,178**	**7,727**	**11,618**	13
Colombia	A	13,944	9,814	10,919	9,995	9,177	10.2
Côte d'Ivoire	R	4,283	2,494	4,713	2,406	6,110	6.8
Indonesia	R/A	6,903	3,947	5,755	5,115	5,358	6

Source: International Coffee Organization (2015)
[a]*A* Arabica, *R* Robusta

Canada 2002). Since 2000, Vietnam has emerged as the second largest coffee producer in the world behind Brazil (from 16th in 1990 to 6th in 1995) (ibid) (Table 2.1).

Coffee has been mainly planted in the Central Highlands. In 2001, the total coffee area of Vietnam was 566,800 ha, of which the Central Highland made up 476,800 ha, and accounted for 72% of the coffee area and 92% of the coffee production in Vietnam (Dang and Shively 2008; Tran 2007). Notably, in 2000 the coffee area in Dak Lak Province alone was 257,100 ha, and accounted for 60% of its arable land, and achieved approximately 55 % of Vietnam's annual Robusta output (Cheesman and Bennett 2005; UNHCR 2006). In this early period, the coffee boom prompted a massive influx of migrants to the Central Highlands; they keenly sought land and developed it into commercial coffee plantations. Coffee beans were elevated as "brown gold," and this gold lured many minorities, who had ever downplayed coffee value and refused to plant coffee (as coffee could not give them food) and who were now eager to join others in planting this "miracle crop" (UNHCR 2006). Small farmers were encouraged to expand their coffee gardens. "Monoculture coffee planting has been the most widespread" (Dang et al. 2001). Household incomes in the Central Highlands in 1999 mainly came from profit of coffee, and the contribution of coffee-derived income was 87%.The average annual household incomes in Dak Lak were USD204 per capita in 1990, which then reached USD390 per capita in 2000 (Dang and Shively 2008).

However, this prosperous period did not last long. From early 1999, the world fell into an acute coffee crisis, and the coffee price dropped sharply. That threatened small coffee farmers around the world and put economic growth and social stability in coffee-producing countries at risk. Vietnam and its Central Highlands was no exception. Many poor and unskilled farmers went bankrupt. The global coffee crisis has been blamed on the expansion of the coffee supplied by nontraditional suppliers, especially Vietnam,[3] which created a glut of coffee that drove coffee prices to

[3] In pursuit of the very attractive prices of the mid-1990s, the Vietnamese government promulgated favorable policies regarding coffee production. Vietnam then succeeded in boosting coffee production at times of high prices. However, the expansion of coffee plantations exceeded the plans, and coffee was planted in areas which had not been designated for coffee. The coffee boom in Vietnam did not follow market trends, and they exported increasing quantities of coffee to the world market,

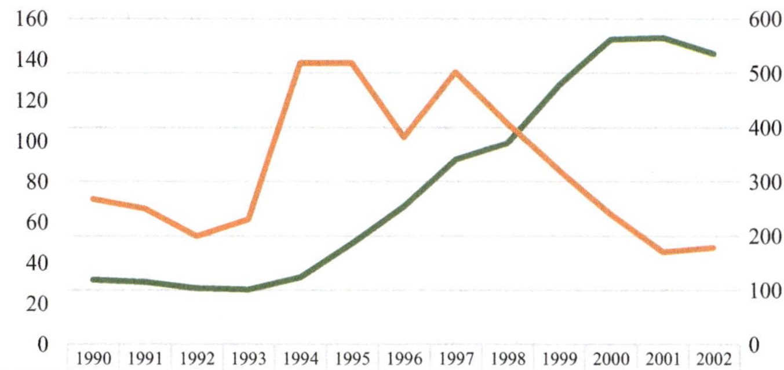

	1990	1991	1992	1993	1994	1995	1996	1997	1998	1999	2000	2001	2002
Vietnam coffee area (1000 ha)	119.3	115.1	103.7	101.3	123.9	186.4	254.2	340.3	370.6	477.7	561.9	565	535.5
Coffee price (US cents/lb)	71.53	66.8	53.35	61.63	138.45	138.42	102.07	133.91	108.95	85.71	64.24	45.59	47.74

Fig. 2.2 Coffee prices and Vietnamese coffee area (1990–2002) (*Source*: Tran 2007; General Statistical Office 2010; International Coffee Organization 2015, drawn by author http://www.ico.org/about_statistics.asp, drawn by author)

the historic low. Vietnam was also blamed for the increase in the low-quality coffee market, when the Vietnamese coffee crops were mainly "Robusta", a less flavorful coffee than the traditional Arabica coffee (White Paper n.d.).

During this crisis period, the profit from coffee was insufficient for farmers to cover the variable costs of coffee production (Dang and Shively 2008). Understandably, the most acutely affected area was the largest coffee-producing area in Vietnam: the Central Highlands. Coffee had been the main factor in the economic growth for this area (Fig. 2.2).

Consequences of Coffee Market Change to Land and Indigenous People

During this time, migrants both bought land from indigenes and hired the indigenous men with knowledge of the forests to clear the land for coffee. Some simply started clearing land or claimed the already cleared land, some of which was the old swidden land or community fields. Less than 4% of migrant households that came to this upland received land from the state; 47% purchased privately, and 46% seized through clearing and preparing "unclaimed" forest land (MOLISA 1997, cited by McElwee 2001). The expansion of the agricultural area contributed to the deforestation of a place that had always been known as the richest and most densely forested area in Vietnam. Cutting trees and deforestation for farming became an alarming problem in the Central Highlands. Deforestation, and intensive irrigation, also led to soil erosion and serious water shortages (Duncan n.d.).

The coffee crisis in the late 1990s made this situation worse, since the farmers neither gained profit from coffee nor built up enough capital to invest. In 2001, in an effort to balance global supply and demand, a plan was announced for key coffee

without any attention to pricing. The overproduction of coffee was exacerbated by the existence of quantities of defective and low-quality coffee beans in the global market (Ponte 2002; Duncan n.d.).

growers in Dak Lak and Lam Dong to cut a total of 110,000 ha of coffee trees to plant cocoa, cotton, or maize. In Vietnam, 250,000 ha of coffee was projected to drop between 2000 and 2005 (Human Rights Watch 2002).

The indigenous people did not profit much from coffee, even during the high price period. This was of course much worse during the coffee crisis. The minority farmers did not profit as much as the Kinh people, because they had less access to the technical assistance, credit, and the market. In 2003, while the Kinh people achieved a significantly high income per capita of USD 7.6 million, the indigenous people achieve a modest income of USD 4.4 million (Dang and Shively 2008).

The rapid decline in coffee prices put the economy of the Central Highlands at risk; however, this led to different responses from the indigenous and Kinh farmers. The capital-endowed Kinh households were better equipped to deal with the fluctuation, as they had managed to store coffee with an expectation of regaining price or reduced their investments (applied minimal amounts of fertilizer, stopped investing in inputs). The Kinh people were also more flexible in turning to other sources of income. When researching the reactions of households in the Central Highlands to the collapse in coffee price, Dang and Shively found the following to be the most common household responses: no response, input reductions, changed crops, and enhanced liquidity (either by borrowing money, seeking off-farm employment, or selling livestock) (Dang and Shively 2008). The options of the indigenous people were more limited. They appeared to have very few opportunities to enhance liquidity, whereas their Kinh counterparts had better access to credit and off-farm work. The indigenous people owned land as their unique asset and heavily depended on agricultural as their main income; thus, they suffered more. Since their key response was not to supplement their incomes in other ways, they were required to cope with the price plunge in more fundamental ways (Dang and Shively 2008). Some of them were forced to transfer or lease a part of their lands to pay for the debt and to continuously invest their coffee. In some cases, this led to people losing their land forever, pushing them further into the forest to return to subsistence swidden cultivation. However, given the expansion of agriculture and the population boom, there was no longer enough land to practice the rotational shifting cultivation system. Some had no choice other than to neglect their gardens, exposing them to drought, and even cut down their coffee trees (UNHCR 2006). Many hectares of freshly planted coffee trees were cut down, and the land was then planted with other crops.

2.1.3 Population Trends

Another trend that changed the face of the Central Highlands is the population growth. State-organized migration programs along with spontaneous migration of lowlanders to the Central Highlands brought about a transformation in these highlands.

Table 2.2 Evolution of population of major region in Vietnam (1926–1991)

Region		1926		1991	
		Total	% of the total	Total	% of the total
I	Northern foothills and mountains	1,590,757	9.39	11,836,000	17.41
II	Red River Delta	5,713,242	33.72	13,518,000	19.88
III	North Central Coast	2,843,139	16.78	9,287,000	13.66
IV	South Central Coast	2,324,844	13.72	6,922,000	10.18
V	**Central Highlands**	**423,420**	**2.50**	**2,804,000**	**4.12**
VI	Northeast Cochinchina	1,340,330	7.91	8,405,000	12.36
VII	Mekong River Delta	2,705,689	15.97	15,222,000	22.39
Total		16,943,678	100.00	67,994,000	100.00

Source: De Koninck and Déry (1997)

Population Growth

The plan to resettle people in the Central Highlands, and relax the high population density in lowland, was highly supported by the trends in the coffee market. Both the government migration policy and the coffee market pushed the population growth exceeding the initiate plan and brought about the dramatically change in the uplands society (Table 2.2).

When considering the historical background of the Central Highlands and Vietnam after the unification, and the wide disparity in population density between these highlands and the other regions, the transmigration was an understandable and indispensable trend. However, in reality, this resettlement process happened at an extraordinary rate and took on enormous proportion, as described by De Koninck and Déry (1997, p.11): "The magnitude of the transformation of the relative demographic weight of the various regions of the country probably has no equivalent in the region."

Central Government Program

The settlement program was formulated and issued through Resolution No.95-CP, 1980 (Thu vien phap luat 2012). According to the central government program, many areas in the post-reunification highlands were designated new economic zones (NEZ – *vùng kinh tế mới),* where people from the densely populated provinces and other minorities from the northern highlands were encouraged to move with preferential policies, friendly pressures, and even outright force (McElwee 2008).

According to the Second Five Year Plan (1976–1980), the state intended to relocate 10 million people to this area by the end of the twentieth century. In the first period, 4 million people would be resettled: 1.5 million residents from the large southern urban centers and 2.5 million from the overcrowded northern areas (UNHCR 2006). The greatest aim of the NEZ program was to populate the Central Highlands (McElwee 2008). The resettlement was accompanied by the sedentarization of 700,000 central highlanders (UNHCR 2006) through the Fixed Cultivation and Sedentarization program. According to this program, the villagers were sedentarized in new villages for a stable and more developed life. However, whereas the migrants were enthusiastic to participate, the indigenous people avoided joining. Although the program allowed indigenes access to the allocated land, that land was not enough to continue their tradition of shifting cultivation. And although the government funded them from 100 to 135 billion VND (US 8.5–10 million) per year, this was not enough for them to set up permanent cultivation (McElwee 2008).

Until 1976 the population in the Central Highlands was 1,225,000 people with a population density of approximately 21.96 people/km^2; meanwhile, the average population density in Vietnam was 148 people/km^2 and the Red River Delta areas exceeded 1,200 people/km^2 (Le and Pham 2014). Ten years after the reunification (1976–1986), the population in the Central Highlands increased to 2,013,000 people, making the growth rate in these 10 years 6.43%. In the next period, the growth rate slowed down but still increased by 5.9%, and the population rose by 1,086,000 by the early mid-1990s (Fig. 2.3).

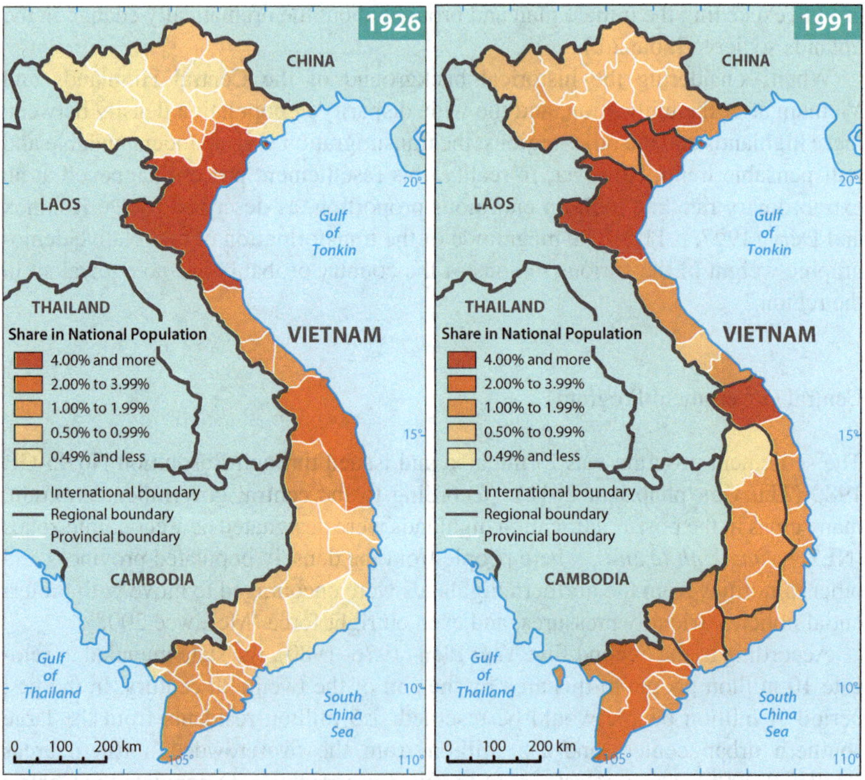

Fig. 2.3 Vietnam: Relative weight of provincial population in 1921 and 1991 (*Source:* De Koninck and Déry 1997, edited by Volker Schniepp)

Unplanned Migration

Following the organized resettlement, the abolishment of travel restrictions in reform policies, and the economic motivation in response to the increased international demand for coffee, a large number of spontaneous migrants also flocked to the Central Highlands in search of their fortune from coffee growing.

The rocketing price of coffee generated a population boom: from a low of US 53.35 cents/lb in 1992, the average annual composite indicator price for coffee beans increased to a high of US 138.42 cents/lb in 1995; and the population influx to these uplands exceeded planned migration by a factor of approximately 12:1 between 1995 and 1991 (Cheesman and Bennett 2005; McElwee 2001). In the last 5 years of the 1990s, the population grew from 3,099,000 to 4,059,928 with a growth rate of 7.75%.

According to the recommendation of the International Labor Organization, the population pressure threshold should not exceed 3%. The population growth rate of the Central Highlands, however, doubles the recommended percentage (Le and Pham 2014). The population density increased more than fivefold in those four decades, from 17 people/km^2 in 1976 to more than 90 people/km^2 in 2002 (Dang and Shively 2008). Overall, between 1976 and 1999, the population of the four provinces of the Central Highlands increase by 231% which added 2,834,928 people to this area (Nguyen 2002), with the indigenous population growing from 600,000 to approximately 1,000,000 (MOLISA. 1997 cited by McElwee 2008). It is also important to take into consideration that the statistics in this region may represent underestimation (Fig. 2.4).

Dak Lak was the province that attracted the most migrants due to its vast suitable land for perennial crops. The statistical data for this province reveals how spontaneous migrants contributed to accelerating the organized resettlement and even

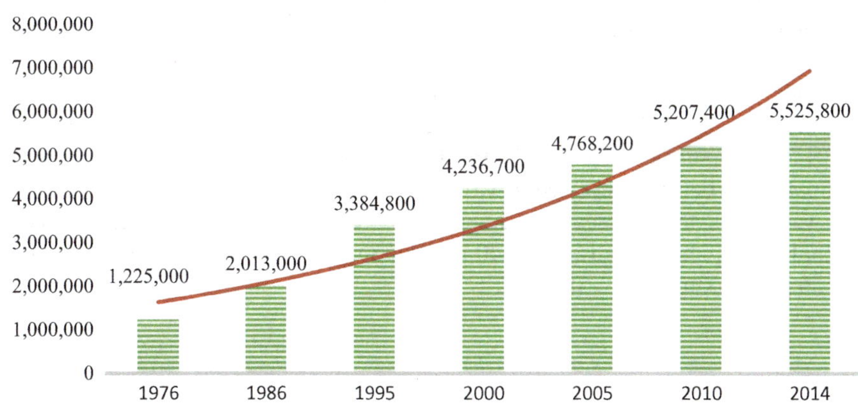

Fig. 2.4 Population of Central Highlands (1976–2012) (*Source*: Hardy 2005; General Statistical Office 2015, drawn by author)

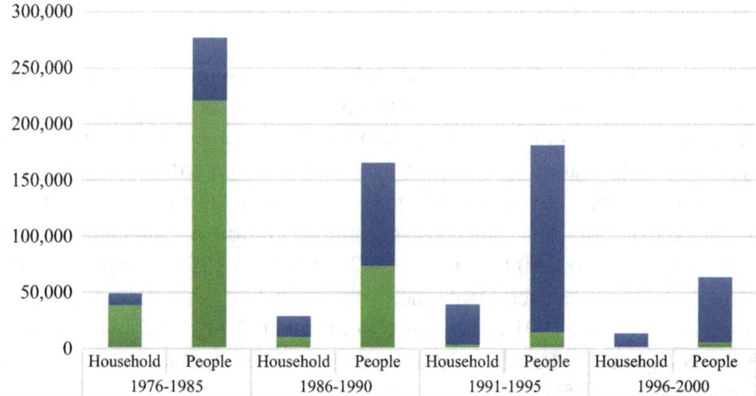

	1976-1985 Household	1976-1985 People	1986-1990 Household	1986-1990 People	1991-1995 Household	1991-1995 People	1996-2000 Household	1996-2000 People
■ Spontaneuous migration	10,258	55,390	18,338	91,658	35,580	166,227	11,828	57,827
■ Organized migration	38,347	220,871	10,131	73,242	3,301	14,334	1,202	5,382

Fig. 2.5 Number of migrants to Dak Lak (1976–2000) (*Source*: Nguyen 2002, drawn by author)

exceeding this plan. The skyrocketing of migration in Dak Lak at that time partly reflected immigration status in the Central Highlands.

Between 1976 and 1996, Dak Lak received a number of around 311,000 planned NEZ migrants and an additional 350,000 spontaneous migrants in the same interval (UNHCR 2006). In the late 1990s, the spontaneous migrants constituted the majority of the population growth in this province. In 2001, the population in Dak Lak reached 1,950,000 with approximately 1,500,000 of these people living rurally (Cheesman and Bennett 2005). Overall, between 1976 and 2000, 128,985 households migrated to Dak Lak, while just 53,370 households migrated to Lam Dong, 18,000 households to Gia Lai, and 3,600 households to Kontum (Nguyen 2002) (Fig. 2.5).

The migration trend of the Central Highlands from the 1990s to now has undergone a significant change, with the volume and the rate now slowing down. The immigration rate in 2009 was a half of what it was in 1999, and the ex-migration rate increased twofold in the same interval. That resulted in a sharp drop of the net migration rate from 70.02% in 1999 to 12.8% in 2014. Although the population growth rate dropped, this number is still higher than the national average rate (10.3% in 2014). The population in the Central Highlands reached 5,525,800 people, and the population density increased to 101 people/km2 in 2014. However, the population was distributed differentially among the different provinces. Dak Lak gained the highest population density of 140 people/km2, whereas this figure in Kontum was 50 people/km^2 (Fig. 2.6).

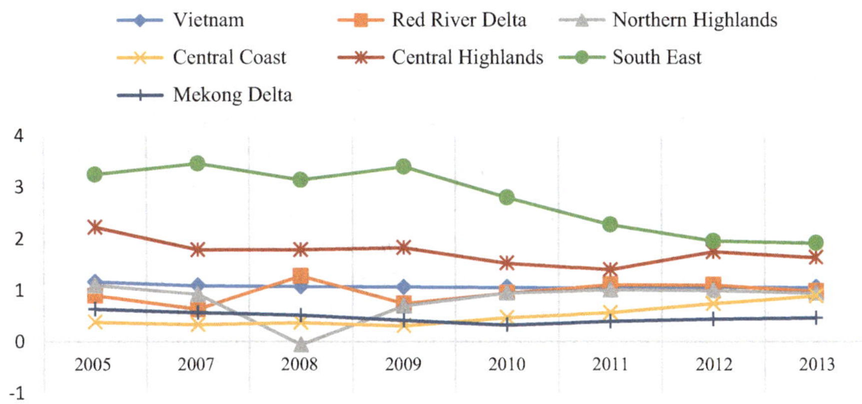

Fig. 2.6 Population growth rate of region in Vietnam (2005–2013) (*Source:* General Statistical Office 2015, drawn by author)

Impacts of Population Trend on Socioeconomic Status in the Central Highlands

The population has boomed at an unprecedented rate over the last decades and has generated a substantive impact and brought a number of challenges to the uplands' socioeconomic development. In particular, the free migration has presented major administration problems. The authorities did not recognize the presence of spontaneous migrants for some months (Hardy 2005), even after they had already cleared land and destroyed forest for their cash crops. Without state's sponsorship, spontaneous migrants coped with difficulties in starting their new lives. Although the southeast was also experiencing a high growth rate at the same time, that region was a leading industrial center with ample infrastructure and facilities. Meanwhile, the Central Highlands was originally a vast area of secluded and pristine forest.

The massive migration was an overburden on this area in terms of the environment, social security, housing management, employment, and urban services. Some migrants had to search deep into forests to find land to cultivate and unofficially settled there, accepting the lack of communication services, infrastructure, and facilities. In order to address the negative impacts of the migration, a number of conferences have been organized to discuss solutions and find ways to restrict immigration. More notably was the national workshop on the issue of internal migration in 1998 in Hanoi and the recent conference "Định hướng giải pháp về vấn đề dân số, di dân trong phát triển bền vững Tây Nguyên giai đoạn 2011–2020" (Orienting solutions to population, migration issues for sustainable development of the Central Highlands 2011–2020) (Institute of Sociology 2014). This spontaneous migration also aroused the concern of some international organizations such as UNFPA.

Change in Population Composition

Concurrent with the population boom, the population composition also changed rapidly in the Central Highlands. The indigenous population is racially and linguistically distinct from the lowland Vietnamese (Hickey 1967). These indigenous groups have undergone many name changes. In precolonial times, they were called *Moi* by lowland Vietnamese. In early colonial times, the French labeled them *savages*. Then, they were referred to as *Montagnard* (literally "mountain dwellers") which was also used by Americans after 1955. An American anthropologist translated this into highlanders. The South Vietnamese regime labeled them as *highland compatriots (đồng bào thượng)*. Finally, the Vietnamese government now prefers *ethnic minorities (dân tộc thiểu số)*. This means that the indigenous population of the Central Highlands no longer has a generic label; instead, a simple distinction is made between the *Việt* ethnic majority and the 53 officially recognized ethnic minorities (UNHCR 2002).

Indigenous people comprise multiethnic groups. Although these groups vary in their ethnic backgrounds, they share some similar specific features of traditional economic systems and social organizations and are also exposed to the same transformation after Vietnam War. More importantly, the diversities of socioeconomic situations between three study sites are more emphasized in this study than the differences of ethnic backgrounds. Therefore, the term of indigenous people is used to generalize these ethnic communities who have lived in this area since precolonial times and to distinguish them from other migrant groups. Since 1954, some North Vietnamese highlander groups have also settled in the Central Highlands, including the Tay, Nung, Thai, Muong, and Yao (Hickey 1967). These ethnic groups are not part of this study. According to the "Nomenclature of Vietnamese Ethnic Groups" issued in 1979 (No. 121-TCTK/PPCD), the indigenous population in the Central Highlands comprises 12 groups: Gia Rai, E De, Chu Ru, Ra Glai (belonging to the Malayo-Polynesian language family) and Ba Na, Xe Dang, Je Rieng, Brau, Ro Mam, Ma, M'Nong, and K'Ho (belonging to the Mon-Khmer language family) (Le and Pham 2014; McElwee 2008).

Until the start of the mid-nineteenth century, the overwhelming majority of the Central Highlands was made up of highlanders. Since the end of the nineteenth century, lowland people began appearing in this area as a result of commercial and political contacts. Highlanders, however, maintained an autonomous social and cultural space until this century (UNHCR 2002). The share of ethnic Vietnamese at that time represented a very modest proportion. The demographic diversification only intensified after 1975 reunification (De Koninck and Déry 1997).

Since the 1990s, the indigenous people have become the minorities in their own land (Hardy 2005), while the number of ethnic Vietnamese has continuously increased. From 1953 to 1989, the Kinh population increased from 30,000 to almost 2,000,000 and accounted for about two thirds of the population of these uplands (UNHCR 2006). In Dak Lak, the largest province in the Central Highlands, the proportion of Kinh was about 70% by 2000. The total population in this province was 1.8 million, of which 400,000 were ethnic minorities. In contrast, the Kontum

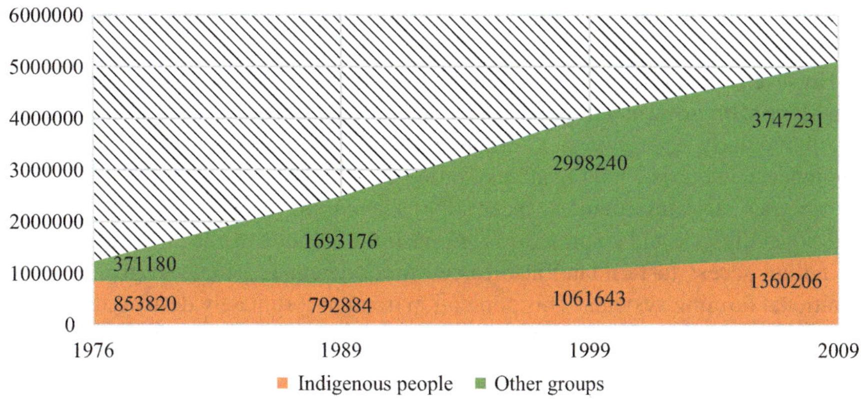

Fig. 2.7 Change in ethnic composition of the Central Highlands (1976–2009) Source: (General Office for Population Family Planning 2010; Le and Pham 2014), drawn by author

province still retains a majority population of minorities, who accounted for 53% of the population in this province in 2009 (Kontum Province 2012) (Fig. 2.7).

One thing that should be emphasized is that the influx of Kinh migrants was also followed by other ethnic groups from the northern highlands "who were fleeing overpopulated areas and lack of land in their home provinces" (McElwee 2001, p.3). "Up to the mid-1990s, official statements tended to characterize free migrants to the Central Highlands as ethnic minority" (Hardy 2005, p.310). From 19 ethnic groups in 1954, the population in the Central Highlands became more diversified with up to 40 groups in 1989, and nowadays, most of ethnic groups of Vietnam already present in these south uplands (Le and Pham 2014). That presents social and cultural problems in the Central Highlands. By 2009, the ethnic minority migrants accounted for 8.73% of population in the Central Highlands; of them the Tay accounted for 2.65% and the Nung accounted for 2.05% (idib). The influence of migration was first reflected in the change in ethnic structure (Hardy 2005). Vietnam's Central Highlands, which had always been known as a specific culture area that kept outside cultural influence to a minimum, was now exposed to a situation of extreme cultural diversity. The predominance of the Central Highlanders was gradually being diluted, whereas the ethnic Vietnamese were increasingly pushing their way out of their own traditional lowland domain (De Koninck and Déry 1997).

The population transformation in this area has generated a number of unanticipated results. The most notable thing was the land conflict between the indigenous people and the immigrants. The research by McElwee (2008) presents numbers on the land disparities: indigenous inhabitants had an average area of 0.23 hectare/household, while newcomers had an average land area of 1–2 hectare/household. When comparing the indigenous people and the ethnic minority immigrants, Le and Pham (2014) stated that particularly in Dak Lak and Dak Nong, the land funds of the ethnic migrants were much higher than of the indigenous groups. The NEZs had created competition over land and natural resources, and the massive influx inevitably resulted in land disputes.

However, the consequence was not limited to the above issue. The transmigration along with government policies also led to transformation in highlander' society, particularly in terms of a negative impact on the traditional culture. For highlanders, who had long-standing settled on their ancestral land in traditional long-houses, the resettlement meant the abandoning and destruction of their traditional lifestyle and community. The constraints related to allocated land and the dwindling access to farmland also forced the indigenous people to give up their traditional farming systems. This population trend unfortunately destroyed the customs and in turn weakened highland communities. The population policies related to resettlement and sedentarization should clearly have been carried out with more respect and sensitivity. In November 1989, the Politburo issued Resolution No. 22, which partly admitted to some of the past mistakes in regard to the establishment of NEZs, state farms, and cooperatives in the highland regions and simultaneously appealed for a development strategy on the "basis of respect for local cultures and of the family economy" (Salemink 2003b).

2.2 Land-Use Change: Consequences of Policy Reforms and Market Changes

Policy reforms, market conditions, and population trends jointly brought about the vulnerability context now evident in the direct change in land use and, subsequently, in the indirect influences on the highlanders' livelihoods. In recent research, Moreda (2012) firmly stated that "land still remains a critical resource for developing countries where a vast majority of its population continues to live in rural areas as farmers." Indeed, for the indigenous people in the Central Highlands of Vietnam, land has long been a key productive asset.

2.2.1 The Common Trend of Land Transformations in the Central Highlands

In the more than 40 years since the Vietnamese government applied the policy reforms, the story of this dramatic change in land use is gradually becoming old. Tracing back statistics and land-use change maps from 1975 is beyond the scope of this study. In light of our limited access, the data is gathered from a few decades ago and then forms a recent picture of the land use in this area that is directly impacting indigenous people's life at the moment.

Before 1975, Vietnam's Central Highlands was sparely populated; forest covered about 85% of the total land area (Dang et al. 2001). The rich forest remained more or less untouched, at least until 1960, when exploitation of these uplands began under the regime of the Republic of South Vietnam (Hickey 1967). After the Vietnam War, in order to forestall mass starvation, the government embarked on a program of rapid and intensive agriculture in the Central Highlands. Between 1976 and 1990, the forest lost 30,400 ha per year (Rietbergen-McCracken et al. 2007, p. 52). By 1990, the deforested areas in these uplands extended to 1,400,000 ha, approximately 26% of the total land area. Remarkably, the cleared forest land supporting the program of sedentarization and NEZs alone spanned up to 95,000 ha (Gialai to Kontum Province 52,400 ha, Dak Lak to Dak Nong 34,800 ha, and Lam Dong 7700 ha).

After 1990, Vietnam's government encouraged the local people to exploit the bare land, instead of clearing more forest. Between 1991 and 1995, the rate of forest loss declined to 25,000 ha per year (Rietbergen-McCracken et al. 2007, p. 52). However, not long after that, the Central Highlands witnessed the second drastic land transformation due to coffee market impacts. Around 486,500 ha of land was converted to agricultural use from 1995 to 2000, of which 136,000 ha was forest land and 308,000 ha was unused land. Moreover, 344,700 ha of natural forest was transformed into barren land (Le and Pham 2014, p. 317). The greatest transformation occurred in areas planted with coffee, which nearly doubled the perennial crop area (from 339,056 ha to 652,855 ha) in the period following the coffee boom. Along with the enormous demand on agricultural land, forest land was also sacrificed for the development of urban areas, infrastructure, and other nonagricultural uses under the pressure of the population boom.

Since 2000, there have been no major events that have accelerated the land transformation. However, the trend of land-use change still follows the path of the previous period, as land devoted to agricultural and nonagricultural use (urban or built-up) keeps increasing, with a commensurate loss of forest (Figs. 2.8 and 2.9).

Agricultural areas expanded from 22.6% of the uplands in 2000 to 36.6% in 2014. In spite of the fluctuation of the various markets, particularly the coffee mar-

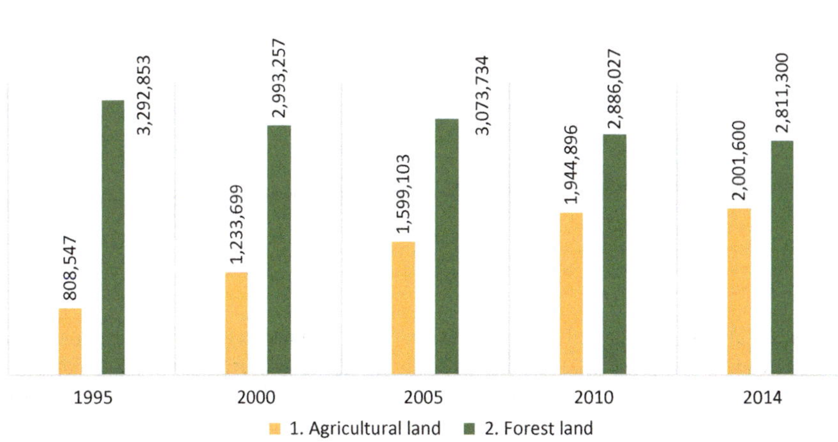

Fig. 2.8 Land transformation trend in Vietnam's Central Highlands (1995–2014) (*Source*: Le and Pham 2014, p.315; General Statistical Office 2015, drawn by author)

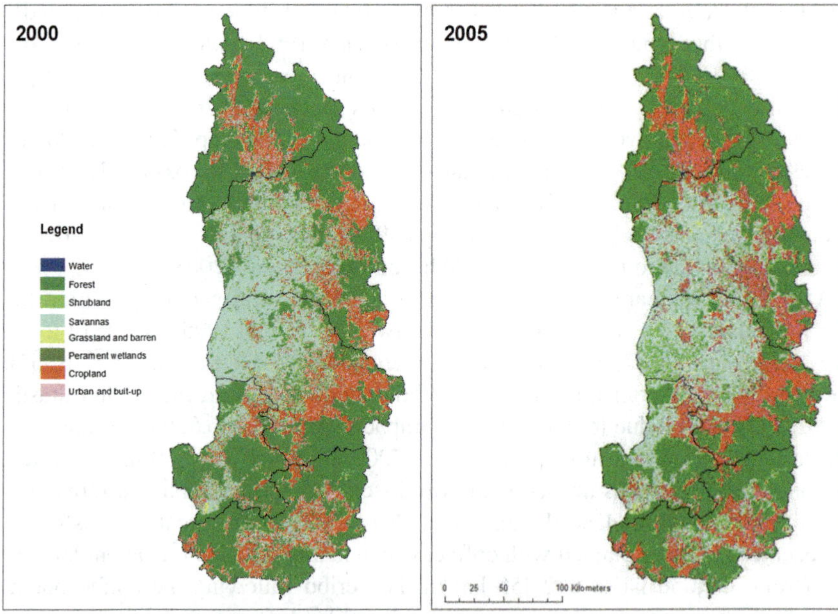

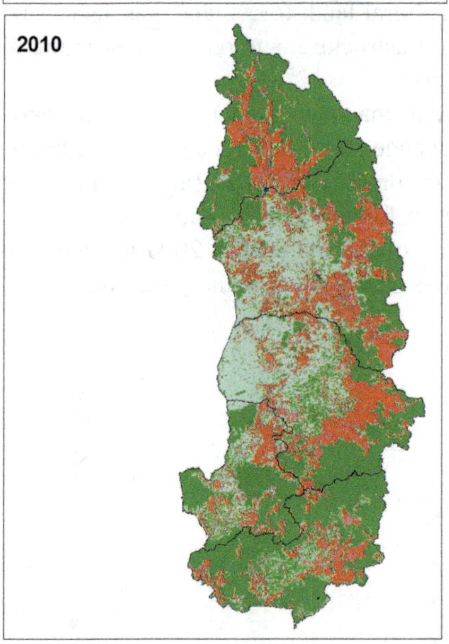

	2000	2005	2010	Change 2000/2010
Forest	2,555,941	2,324,012	2,468,991	-86,950
Cropland	775,611	858,023	998,943	223,332
Shrubland	306,235	285,482	313,206	6,971
Savanna	1,176,869	1,329,619	1,027,001	-149,868
Wetland	177,768	170,950	176,748	-1,020
Grassland/barren	181,932	176,679	168,643	-13,289
Urban	288,185	313,129	302,614	14,429
Water	1,603	1,659	3,407	1,804

Fig. 2.9 Land use-land cover change in Vietnam's Central Highlands (2000–2010) (*Source*: The MODIS Land Cover Type product (short name: MCD12Q1) was retrieved from the online Data Pool, courtesy of Active Archive Center (LP DAAC), USGS/Earth Resources Observation and Science Dakota, https://lpdaac.usgs.gov/data_access/data_pool)

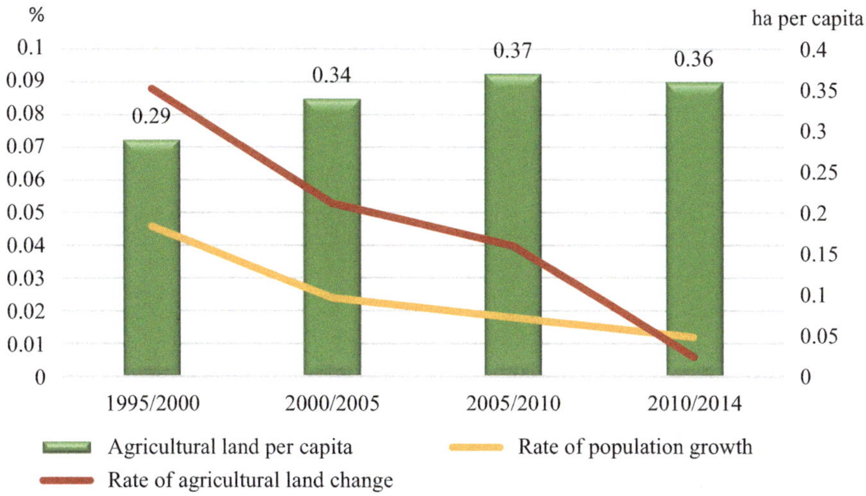

Fig. 2.10 Integration of population and agricultural land transformation in period (1995–2014) (*Source*: General Statistical Office 2015, drawn by author)

ket, the highest expansion still appeared in perennial crops. The perennial crop areas increased by 170% between 2000 and 2010, reaching 1,093,787.3 ha. As a result, the forest area shrank from 54.8% in 2000 to 51.4% in 2014. In addition to the reduction of forest area, forest quality has also fallen. Between 2000 and 2005, all forest types have declined, most notably the evergreen broadleaf forest, which fell rapidly from 2,256,058 to 2,089,998 ha. Dang and his colleagues stated that the remaining forest was mainly of poor quality, located on the tops of the mountains or on very steep and rocky hills, and not suitable for cultivation (Dang et al. 2001). Since 2005, the forest area has increased slightly as a result of forest rehabilitation and plantation programs.

When looking at insights provided about the correlation between trends in population and land-use change, it is clear that the population boom was closely followed by the extension of agricultural land. Within the last decade, the growth rate of both elements has slowed. Recently the rate of land extension has been unable to keep pace with population growth, and in 2014 the rate of land extension dropped below the rate of the population increase. This shows that the land with development potential is running out (Fig. 2.10).

In the past two decades, the growth rate of agricultural land has always been higher than population growth; land area per capita subsequently increased slightly but steadily. And this average in the Central Highlands is rather higher than other regions in Vietnam, at 0.3 ha per capita in comparison with the national average of 0.11 ha per capita. However, this cannot reflect the status of the indigenous people, since they are now minority groups. Ironically, the indigenous people currently inhabit a smaller and less fertile territory, while they depend more heavily on agri-

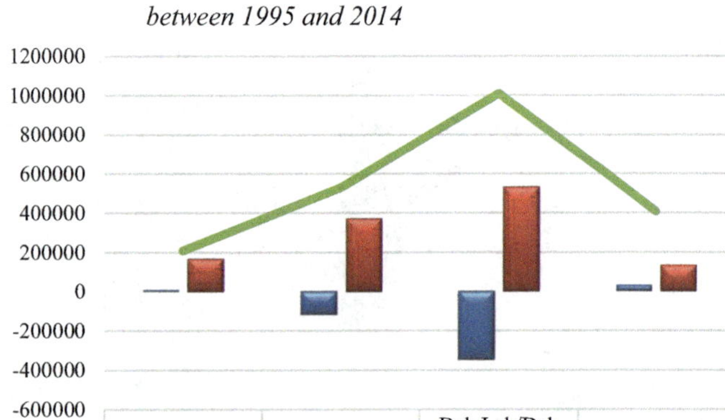

	Kontum	Gia Lai	Dak Lak/Dak Nong	Lam Dong
Change of Forest (ha)	6800	-118500	-348800	28900
Change of Agricultural Land (ha)	164100	368400	528600	132000
Change of Population (people)	204700	527100	1006300	403000

Fig. 2.11 Change of population and land use in Vietnam's Central Highlands between 1995 and 2014 (*Source*: General Statistical Office 2010, 2015, drawn by author)

culture than the Kinh majority (United States Agency for International Development 2008). Although they all might have received an equal share of land after the state-run farms were dismantled, the disparity in territory size has myriad causes.

Again, the correlation between population trend and land use-land cover change can be seen when the comparison among different provinces is investigated. Since Dak Lak[4] was a notable example of skyrocketing migration, this province also saw a drastic change in land use. Between 1995 and 2014, the population in Dak Lak grew by 1,006,300 people; at the same time, the province lost 348.800 ha of forest and gained 528.600 ha of farmland. In contrast, the low population growth in Kontum is accompanied by less change in land use. In the same period, the population of this province increased by 204,700 people, and their agricultural land expanded by 164,100 ha. Different trends of land use over these uplands will be discussed further, as we closely examine three study sites in the following section (Fig. 2.11).

[4]In 2005, Dak Lak was divided into two separate provinces: Dak Lak and Dak Nong (Wikipedia 2014). In order to calculate the changes in land and population between the provinces from 1995 to 2014, the data of Dak Lak and Dak Nong in 2014 are counted together in order to make a valid comparison with 1995.

2.2.2 *Various Land-Use Sceneries at Study Sites: Kontum City, Lak District, and Lac Duong District*

Within the common trend of land transformation in the Central Highlands, there were various land-use sceneries and different productive levels among the three study sites. At the first glance, land use in Kontum City was the most varied and reflects the typical characteristics of an urban area: the residential area was rather large, whereas the forest area was small due to years of exploitation. In contrast, thanks to the existence of the national park and natural reserve, the forest areas in Lak and Lac Duong District[5] were dominant; these two areas also showed a moderate level of urbanization (Fig. 2.12).

In Kontum City, annual crops occupied the largest area of the agricultural land. Recently, the area of perennial crops, specifically rubber, has increased. This land belonged mostly to individual Kinh people or companies. With its higher productive level of using land, Lak District was well-known as the granary of these highlands. Paddies accounted for around 50% of the agricultural land. Most of the fertile land was located around Lak Lake. Annual crop lands were interspersed between the paddy fields. The area of perennial crops, mainly coffee, have expanded quickly following the arrivals of migrants. Another picture was observed in the Lac Duong District. With forests taking up to over 80% of the total area of land, agricultural land in this district was more limited. Perennial crops constituted up to 68% of the agricultural land. A total of 2,624 ha was used to grow annual crops in 2014. And more notably was the existence of a high-tech crop: flower. Lac Duong was recognized as the most productive land use among the study sites (Fig. 2.13).

2.2.3 *The Precarious Situation of the Indigenous People*

The indigenes used to have relatively easy lives and lived off the natural forests and abundant lands. As a consequent of the circumstance in the Central Highlands decades ago, indigenous people have been facing challenges, particularly to their livelihood. In rural areas, there is always an important cause-effect relationship between the dramatic changes in land use and livelihood constraints that occurs in roughly the same time period (Sunderlin and Ba 2005, p.4). Likewise, the specific vulnerability context of the indigenes in the Central Highlands is closely related to their dependence on land, both forests and farmlands. Their major economic activity is agriculture, and they rely on the natural forests for the sustenance.

For centuries, the indigenous people lived in the midst of forests and derived great benefit from them. Nearby forests were preserved for collecting timber and non-timber products such as rattan, bamboo, mushrooms, bamboo shoots, and

[5] While district and city have the same level of administration and fall under the control of the provinces, city plays a more important role as a capital of the province.

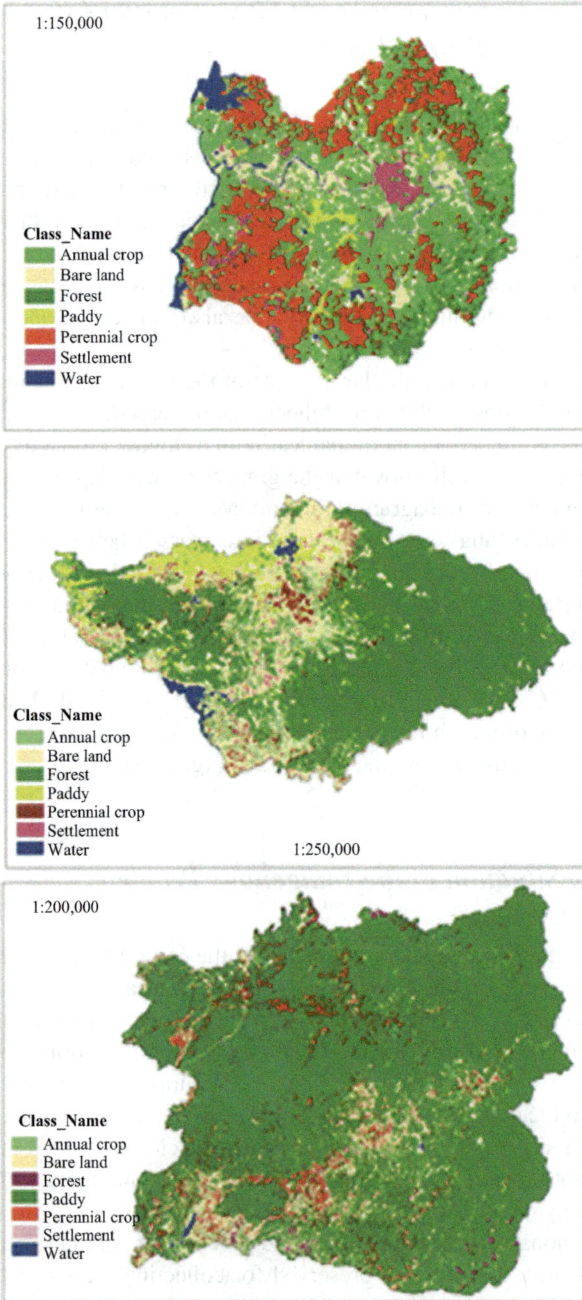

Paddy	1,551
Annual crop	16,465
Perennial crop	11,543
Forest	4,250
Settlement	1,456
Water	1,531
Bare land	6,289
Unclassified	151

Paddy	8,001
Annual crop	4,657
Perennial crop	3,471
Forest	77,651
Settlement	652
Water	1,780
Bare land	29,360

Flower	646
Annual crop	2,624
Perennial crop	6,942
Forest	102,958
Settlement	128
Water	163
Bare land	13,142

Fig. 2.12 Land use in Kontum City, Lak District, and Lac Duong District in 2014

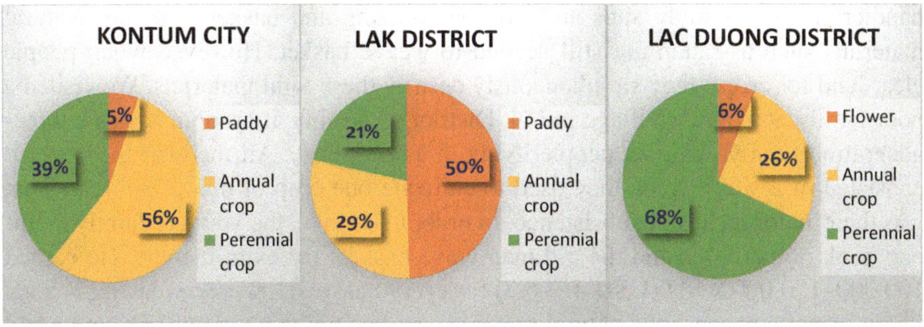

Fig. 2.13 Agricultural land use in Kontum City, Lak District, and Lac Duong District (*Source*: Data is extracted from land-use map in 2014, drawn by author)

medicinal herbs. Today, most of the forest lands have either been destroyed as a result of urban developments such as Kontum City or are managed by the state, and the indigenous people have lost their access to this resource.

Moreover, most of the indigenous people historically engaged in shifting cultivation, and they had a certain plot of forest land for cultivation. Deforestation and the loss of access to what little remained of the forest land has posed a clear threat to the highlanders' lives and livelihoods. Furthermore, in the past highlanders did not cut timber for trade; it was mostly used to build their houses, or less frequently, dugout canoes. Such canoes require a large tree trunk, and it is no longer possible to make them today. The remaining dugout canoes are preserved as part of their traditional heritage and are used in tourism activities. In order to build houses, some indigenous groups ask the Forestry Department for permission to cut timber. However, the fee for such a permission is sometimes higher than they can afford, so the traditional houses are gradually being replaced by brick houses. It requires a great investment for communities to keep their traditional architecture. For example, to renovate the common house at the Kontum study site, every household in the village has to collect and store the natural material for about one year until they have enough materials.

Non-timber forest products were also important for the indigenes' livelihoods. Many interviewees expressed their frustration at the lack of forest resources which had once been abundant. Two such products are firewood and food. At present, firewood has become scarce, and people have to go farther into forest to gather it, or they use wood from industrial crops, instead. Otherwise, they have to purchase expensive types of fuel such as electricity or gas. The loss of food that forests once supplied is also a growing problem. In a self-sufficient economy, many indigenous generations hunted or gathered food from the forests, rather than planting crops or gardens around their houses. Today, the indigenous diet has changed, and poverty is increasing as food becomes more expensive.

The indigenous people in the Central Highlands also have specific handicrafts from which they can gain extra income. Traditionally, these products were made from natural materials which could easily be found in forests. The most popular handicrafts at the study sites are brocade weaving and basket weaving. Natural materials such as rattan are still needed to weave basket. However, when people clear land for crops, they simultaneously destroy these wild materials. We visited a poor family at Nam Pa Village in Lak District on a rainy afternoon. It was a three-generation family, with six people living in a stilt house. All the land they owned consisted of 2 *sào,*[6] which they used to cultivate one crop of maize per year. That was their major means of subsistence. In order to survive, the younger family members took on odd jobs and the old man wove baskets for extra income. He earned 100,000–150,000 VND (USD 4.5–6.7) for every basket. However, weaving is not a stable income, because the natural materials are becoming restricted. This man had to go to Dak Phoi (a commute around 25km from his village) to gather rattan. He was afraid that he *would be unable to find the essential material to continue his work in the future* (Interview at Nam Pa Village, Lak district, on 13 December 2014). Besides, the indigenous people once planted cotton and used wild leaves from the forests to dye the fabric for brocade weaving. Since they do not own the cotton fields anymore, weavers now have to buy the fiber to make their products. This generates a further economic burden. In order to continue producing their handicrafts, indigenes have to invest in the materials in advance, with no certainty of a return on their investment.

In addition to the reduction in forest resources, the influence of changes in land use on highlanders' livelihoods is more obvious when we look at the agriculture sector. Indigenous people have historically depended much more heavily on agriculture than their Kinh counterparts. Farming has thus formed the basis of their economy. The different productivity in land use between indigenes and the Kinh people contributes to explaining why their circumstances at present lag behind those of the Kinh people. Nowadays, the highlanders only cultivate their allocated plots of land to a limited extent. Some of them still grow their traditional crops. The output of these types of crops is low, however, and only provides for daily subsistence or for the purchase of essential goods, which are increasingly commoditized. Their cultivation hardly helps to meet their needs in terms of education, health service, etc. On their limited land, which continues to diminish in size generation after generation, traditional shifting cultivations are virtually impossible. Even grazing cattle, which contributes much to the livelihood mechanism of some indigenous groups, is now more restricted due to the lack of grassland. Indigenes, therefore, have to rotate their crops yearly and try to increase their agricultural productions. However, they still struggle to achieve a relatively comfortable life. Scarcity of land has become an urgent matter among indigenous communities.

Most indigenous households desire to expand their farmland. In particular, cash crops, such as coffee, rubber, etc., are expected to bring a higher income, and indigenous people are lured into following the Kinh in growing these crops. However, reclaiming more land is both illegal and challenging. In order to acquire more land, some try to purchase land. That is however not feasible for most highlanders, as they

[6] Sào: local land measurement unit, equivalent to 1000 square meters or 0.1 hectare in south and central Vietnam

cannot afford the current high prices of land. In addition, even some indigenes who own suitable land, or are able to purchase more land, still struggle to produce cash crops. The unfamiliar cultivation techniques and high initial costs likely make it difficult for indigenes to achieve a high agricultural output. Moreover, market fluctuation is a risk for this vulnerable group. In fact, many highlanders end up having to sell their land to manage their debts, due to the lack of capital and unprofitable farm work.

For now, the livelihoods of the indigenous people of the Central Highlands in general, and at the three study sites in particular, still rely on land. Although the level of development, the way they use land, their specific stories, and the agricultural landscape vary among the study sites, there is clearly a common thread in how the indigenous people struggle with their livelihood constraints and lag behind in comparison to the Kinh majority. If the scarcity of land is not enough to place the indigenous people under stress, the lack of capital to invest in their land in accordance with modern trends of intensive agriculture pushes them into further hardship.

2.3 The Indigenous People's Perception of Their Vulnerability Context

It cannot be denied that the genesis of the current constraint on the indigenous people's livelihood is strongly related to land issues (Moreda 2012). In addition, lessons from the recent development approach have shown that raising the voices and perception of local people will contribute to clarifying their vulnerable context.

2.3.1 Perception of Reasons for Land-Use Change

The indigenous people said that the change in land use was a *common occurrence in the Central Highlands* (interview at Bon Dong Village, Lac Duong District, on 9 November 2014) and pointed out that change levels were various in different parts of the uplands. In fact, the research sites did not witness severe changes in land use since they did not participate in the coffee boom as much as other areas. The change in land use, however, was recognized.

> *Mr. Plyng,*[7] *Bang Gia Village, Lac Duong District*
> *The establishment of NEZs and the massive resettlements happened in all of South Vietnam, not only our district. Even today, the negative impacts of the free migrants have not yet been solved. In our area, however, these events occurred at a more moderate level for political reasons.*

[7] Except for the names of the owners of some specific tourism companies, all names of interviewees are pseudonyms in order to protect the anonymity of participants taking part in this study.

Table 2.3 Planned and free migrations to Lak District (1976–1995)

Migration	Families	People
Planned	1.088	5.115
Free	1.556	8.322
Total	2.644	13.437

Source: Data provided by the Lak people's Committee, interview, May 1996 (Cited by Hardy 2005, p.315)

Highlanders clearly expressed their anger about issues relating to land-use change in the interviews conducted during this study. Understandably, these issues have angered most in Lak District, since this district belongs to Dak Lak Province, which received the biggest number of migrants and where land transformation occurred intensively. The major crop in Lak District is wet rice; however, due to the existence of basaltic soil, which is highly suitable for cash crops, Lak also attracted migrants who wanted to set up their coffee gardens. Notably, the number of ethnic minority migrants in Lak was relatively high. Although the issue has not triggered unrest as it has in some other districts[8] in Dak Lak, land conflict in Lak is still ongoing. One of interviewees told us the following story*: Once I grazed my cow on my former field, which had been already allocated to an immigrant. The new owner shouted at me and told me to stay away. I replied in anger and hopelessness: where is your land? Did you bring it here from the north* (interview at Le Village, Lak District, on 1 Decemeber 2014)? (Table 2.3).

In Lac Duong and Kontum, the conflicts regarding land are, in general, relatively moderate. In addition to migration, most of villagers mentioned the natural population growth within their villages as another reason for the current scarcity of land and the land use changes. After the land reforms, the indigenous people became more restricted in their cultivation activities, because the land they were allocated was much smaller than what they had owned before. Moreover, they blamed the population for land constraint and realized that land could not expand along with the population. Indigenous people had to share a certain amount of their land with the next generations and so on, with result that the amount of land per capita was increasingly decreased. Furthermore, land resources, in indigenes' view, were divided into two types: plain land (*đất ruộng*) and sloping land (*đất rẫy*). While plain land has become restricted, sloping land has been categorized as forest designated as state forests. Indigenous people, therefore, did not officially own the sloping land and were no longer allowed to reclaimed more land either.

[8] In 2001 and again in 2004, thousands of indigenous people in the Central Highlands gathered to demonstrate against government policies regarding religion and land rights (McElwee 2001; UNHCR 2006).

2.3.2 *Perception of Land-Use Change Status*

The scarcity of land has become an urgent matter among the indigenous people. Most interviewees complained that they did not have enough land to cultivate, especially in this modern society.

Mr. Ma Va, M'Lieng Village, Lak District

We were content with the amount of land we had. Our ancestors' land was large, but we only exploited around 1-2 sào for each yearly crop. It was sufficient. We did not expect that life would change so much, and people might need more and more land. The immigrants knew this, so they came and converted waste land into cultivated land. Then they owned this land. They were visionaries, they had money, and they came quickly. What can we do? Some people in my village do not even own any land...

The present study does not intend to gather concrete statistic on the interviewees' land size and transformation, but the amount of land per capita clearly differs among the study sites. On average, it ranges from several *sào* per household in the Lac Duong and Lak District to several hectares per household in Kontum City. However, it is likely that the less land people own, the more productivity they gain from their farm. It is clear that the amount of land owned by highlanders is currently much less than in the past. Even though they achieve a higher agricultural output than before, it is insufficient to support for their lives nowadays. In order to pay for their basic needs, indigenous people need more land and/or need to invest more in their intensive crops. When recalling the past, elders regretfully narrated that *they had lived easily based on nature and land was not high in competition* (interview at Kon Ro Bang Village, Kontum City, December 2014).

According to the perception of the indigenous people, the land constraints have become much worse, since the quality of land and other natural resources are poorer than in the past. The reasons are clearly named as the intensive crops and the adoption of insecticide, pesticides, and chemical fertilizers (so-called plant protect productions). Intensive farming and intercropping were urgently encouraged and promoted by the government after the Vietnam War, since they believed that it was the best way to overcome the threatening mass starvation and to achieve high agricultural output. Indigenous people were forced to produce intensive crops on smaller plots of land. In reality, these government plans gained high results in some regions over Vietnam: agricultural output dramatically increased, and agricultural productions were more highly diversified. However, for the special physical conditions of the highlands, the intensive cropping since the 1990s up to now has caused quicker and worse natural degradation than in the other areas. There has been some research on this situation, which shows that the decrease in natural resource quality has to do with occurrences such as land erosion, degraded watershed function, and deforestation (Dang et al. 2001; Espaldon et al. 2004).

Indigenous people were fully aware of this situation and interpreted it by their indigenous knowledge. In the highlanders' experiences, the recent intensive agricultural systems do not provide the land with enough time to recoup like their traditional farming methods did. In the Lak District, paddy cultivation is the most important type of farming, with the number of crops per year now up to two to three instead just one crop per year as before. Meanwhile, the farmers at the study sites in the Lac Duong District and Kontum City have been farming intercrops, which means they produce different crops each year, from vegetables to annual industrial crops. They have to work harder each year and cannot take time off after harvest time anymore; however, they are still struggling to make a living and cope with poverty. The indigenous people know they can use chemical substances to protect their land and boost their crops, but dependence on and overuse of chemical substances have severe impacts on the environment and people's health; investing in fertilizers and pesticide also pushes the indigenous people into greater risk of debt and poverty.

Mr. Ma Le, Le Village, Lak District

In the past we did not use fertilizers, the paddy fields grew naturally, and we harvested 3 quintals per sào. Every year we produced one single crop in the rainy season, then let the land dry up during the dry season, and the land recovered by itself, because it was so fertile. Since the land reforms, we have learnt to apply technology, and use fertilizers and pesticides. The first season we used these, we gained one ton per sào. However, it did not last long. Year by year the output decreased and has now dropped to 3 quintals per sào like in the past. The land has become "exhausted" now, but we cannot stop using chemical substances, otherwise the output will be even lower. Because of our dependence on fertilizers and pesticides, we now go further into debt with every crop... Not just that, we also think that when we use pesticides and insecticides, they destroy the other nature resources. These chemical substances somehow accumulate in the lake. Then the fish die, and so does the seaweed.

Finally yet importantly, highlanders also acknowledged the change in land use due to deforestation. They were able to explain the local changes that have occurred in nature because of deforestation.

Husband of Mrs. H. Lynh, Le Village, Lak District

People have exploited and destroyed the forest. I am saddened by this. The forests had always been very rich and full of big trees. Whenever it rained, it did not flood. Nowadays, when it rains, the water level rises so high and quickly, and in contrast, when the dry season comes, the land dries up. In the past, it was no yet dry this time of the year [in December]. The lake was full of seaweed, and it looked so green. The fishes were abundant. In the evenings, we could easily fish in the lake and catch ourselves a delicious dinner... People are aware of this negative impact, but they feel that their hands are tied, because they need the land to farm.

2.3.3 Perception of Land Tenure

Although indigenes were clearly aware of land-use changes, many of them were still confused about what a land tenure certificate was. They got used to managing land according to their customary laws, and the land was never officially allocated to them. When the state started issuing land rights with land tenure certificates (Red Book), the indigenous people were expected to learn about this new system and how to manage their land according to this new system. However, land rights are perceived and applied differently in the different communities. At the beginning, highlanders used their land titles incorrectly. As they were not accustomed to the concept of individual land, they were confused by the land tenure certificates and struggled to manage their allocated land which became commoditized. They underestimated the value of their land titles and were careless when conducted transactions relating to their land. As a result, there were immigrants who took advantage of the indigenous people' ignorance and practically grabbed land from them or offered things of little value in exchange.

Most of indigenous people now are accustomed to the land right system and have been forced agree with the idea of the state management of land. They are aware of their limited rights to exploit and benefit from the forests. They are afraid of being fined when illegally exploiting forests. And the Department of Forest Management and rangers act as effective barriers keeping them away from their former forest lands. Also, since they have learned about the value of land rights, indigenes are now more reserved and careful in how they use those rights, especially the right to sell land. They now know that once they sell their land, it means they lose their right to occupy and use the land and that they are unlikely to be able to afford to buy it back. Nowadays, highlanders prefer to lease their land rather than to sell it. However, indigenous people still suffer great losses in their transactions with the Kinh majority, as the lease terms they agree on are often too long and the lease price is too low.

In some cases, communities refuse to give up their ancestral land and do not recognize the state's management of that land. This derives from the mismatching between the new land tenure and customary communal ownership of land. "The land law only cover the permanent agriculture and not swidden plots left fallow" (Human Rights Watch 2002, pp.43). These indigenous people now cultivate permanent agriculture on their allocated land. Besides, they still claim that the fallow land that had been their ancestor land, now is managed by rangers, is their property.

The last case is more complicated, showing that the highlanders intentionally separate the concept of land owner and owner of the land right certificate (red book). In this case, they downplay the official law and insist on maintaining their communal land right. The official land right system is only implemented when the indigenous people carry out business transactions on their land. Buyers are then required to pay twice, once for the land itself and once for its certificate.

Mr. A.lee, Kon Ro Bang I Village, Kontum City

The period of resettlement and de-collectivization have brought disordered to our villages. The authorities allocated my family a plot of land in a neighbor village – the Kon Ro Bang II, along with a land-right certificate. The people in Kon Ro Bang II disputed this and reclaimed their land. Since we are all indigenous people, we knew and understood the customary law. Thus we could not handle this dispute basing on a certificate, like the way the Kinh people do. We simply thought that this land belonged to them, they claimed it, and we gave it back. That's all. After that, my neighbor sold this plot of land to the Kinh people. Since they did not have the Red Book which I was holding, the buyer was able to get it for a bargain price, and then he asked me to sell him my Red Book…

This case clearly reflects how the customary law still holds a certain value in land management. Indigenous people apply land titles in different ways with different counterparts. When land disputes occur among indigenous communities, they follow their traditional law. In another cases, however, as in dealings with immigrants, they accept the land tenure, albeit reluctantly. This leads to dissatisfaction and conflicts. In some cases, the authorities have to intervene and reallocate the land. The differences, misunderstanding, and lack of understanding about land rights make the indigenous people more vulnerable and create difficulties for state management.

2.3.4 Relationship of the Indigenous People with Land and Agriculture

Despite the scarcity of land and the degradation of this resource, the indigenous people insist on sticking agriculture as their major livelihood and are reluctant to practice other livelihoods. However, the reality is that all lands are now either owned by private individuals or belong to state and indigenous people are neither allowed to reclaim the land, nor do they have enough money to buy land. Their deep desire to continue farming strongly reflects their close-knit relationship with land and agriculture. Land constraints have led to highlanders ending up in straitened circumstance.

Mr. Ma Le, Le Village, Lak District

All I have even known is agriculture, nothing more. Everything I have is derived from my farming. My life depends on rice. Unfortunately, the price of rice has now fallen. Still, I prefer to stick to my farming work, and nothing else. I even dream that I can find a fallow field. I keep this dream day after day and try to look for any available land but not successfully. Nowadays, every plot of land already has an owner.

It is clear that land plays the core role in the economic activities of the indigenous people. They firmly state that "farming and lands are their life." The relationship that indigenous people have with lands goes beyond an economic relationship, since they had long used customary law to manage lands, and lands were seen as the historical property which was transferred from generation to generation.

Mr. Ma Va, M'Lieng Village, Lak District

We will never abandon our land and agriculture. How can we? I cannot image it. Ever since the time of our ancestors, our lives have depended on lands, we have become used to it being that way. Indigenous people here never wanted to sell their land. Some are on their beam ends, they are in financial strait; they are forced to sell their land, and then they regret it so much. It breaks their hearts. But there is no other way.

The close tie between highlanders and lands is also evident through the way they exploit and conserve lands. As mentioned above, the indigenous people practiced a traditional agricultural technique, called swidden agriculture, or shifting cultivation, which has long been regarded by many in Vietnam as backward and inefficient but is also known by foreign researchers as luxury cultivation. Under this system, indigenous people left their initial land when it became exhausted to let it revive, then moved to other area of their lands, cleared that land, and continued cultivating. When harvesting forest resource, they only exploited a certain area, enough to meet their needs, and not log timbers for trade. According to their traditional custom, natural resources, especially the forest, have a spirit. Indigenous people, thus, highly respected their resources, and a ritual was always required whenever villagers wanted to exploited resources for their needs. A contrary picture can be observed in the case of immigrants. In the eyes of the indigenous people, the new comers thirstily grabbed land and aggressively cleared it for their crops.

Husband of Mrs. H'Lynh, Le Village, Lak District

Immigrants, mostly Mong[9] *people, have exploited the land so cruelly. They did not have any permission of residence, and seem not to need it at all. Crowds of them flocked into our village, cleared land and set up their farms, so quickly before the authorities discovered their presence. They began exploiting the more remote areas, we hardly noticed what they were doing. The rangers could not stop them. When the rangers blocked one way, they ran and continued exploiting land elsewhere. Then we had no way to send them away. They boldly declared that they would stay here permanently... Some have just arrived recently, and have already established a proper new life with a house, motorbike... This is the same for the Tay people. They occupied land, worked hard, and have become richer than us.*

It seems that highlanders conserve their lands and other natural resources well if they feel these resources belong to them and their ancestors. That has been proven: after the government seized their land, people did not conserve the land and forest like before. Some government programs clearly show that purposely managing and re-greening forests has been not effective, as the indigenous people are only given an employee role in this affair, and they do not feel responsible for protecting and conserving these resources.

2.4 Conclusion

This chapter has described the context of this study and laid a foundation for the subsequent chapters. We employed the approach of double exposure to analyze the two main external causes of local people's vulnerability context: socioeconomic pressures and the natural degradation. More specifically, the exposure context is constituted by policies, population trend, market change, and land-use change. Most of these elements emerged several decades ago, and their influence still persists. This chapter described the historical events based on different sources: historical digital data (satellite images, maps), secondary data, and the historical information from interviewees. These sources cannot be viewed in isolation, as they complement each other, and together give clear insights into the changes that have occurred in the Central Highlands.

Policy reforms are seen as the initial trigger for the transformation in the Central Highlands and paved the way for the remaining factors. The policy reforms led to planned migration, land-use change, land tenure transformation, and even the emergence of a state-oriented market economy. However, this policy system was not able

[9] Mong (H'Mong) and Tay are ethnic minorities who originally lived in northern mountainous area of Vietnam.

to control these above trends, and, in turn, these led to the spontaneous migration, the coffee boom even on unsuitable land, the deforestation, the land erosion, and so on. The vulnerability context of this study cannot be blamed on any single factor, but instead, it is a result of various aggregated factors. The influence of these factors has been objectively analyzed in terms of both advantages and disadvantages. Unfortunately, for the indigenes in the Central Highlands, the disadvantages seem to outweigh the advantages. As such, the indigenous communities have been defined as vulnerability groups, and their circumstance is understood as the vulnerability context in this study.

The livelihood context is not intended to conceptualize as a specific vulnerability in this study, yet, it cannot be denied that livelihood constraint is one of the main stresses that indigenous people in the Central Highlands have undergone. Livelihood has therefore been appointed as a specific aim of this study. Also, the different factors mentioned are not weighed, although land-use change has been highlighted. Land use is acknowledged as being the consequence of a correlation of factors that directly affect indigenes' livelihood. Land use has been investigated at two levels: regional level (the Central Highlands) and at the study sites (Kontum City, Lak District, and Lac Duong District). We obtained a common picture of land-use change over the past decades at the region level and evaluated different contemporary sceneries at three study sites. The common picture of land use clearly depicts the consequence of historical factors, and the different contemporary scenes reflect the dynamic livelihood at the study sites.

In the relationship between land and indigenes' livelihood, it is not easy to pinpoint what is exactly the cause and what is the consequence. However, as the initial plan led to change in land use, the indigenes likely fell into a passive position as the result. Their livelihood decisions are then viewed more as an adaptation, a reaction rather than an intention. In this case, land-use change was caused by external conditions, which led to indigenes being put in a vulnerable position.

Although indigenes are seen as victims in this context, the next chapter will view their role more in light of their active part. The indigenes' perception of their vulnerability context, mainly relating to land-use change, is preliminarily discussed in this chapter. Interventions related to the perception are useful since the perception is recognized as a way to transfer external conditions into certain actions. In this line, a remarkable outcome is the close-knit relationship between indigenes and land, and they are acutely aware of the constraint of this resource. This will be illustrated again in the next chapter when we look into their livelihood typology. And the question *to what extent land-use change influences indigenes' livelihood transformation and whether land-use change is the main means leading to the dynamics of livelihood pathways* should be further examined in subsequent parts.

References

Cheesman J, Bennett J (2005) Managing groundwater access in the Central Highlands (Tay Nguyen), Vietnam. Australia Pacific School of Economics and Government, The Australian National University, Canberra

Coffee Association Canada (2002) The current coffee crisis – cause, analysis, and solution

Dang TH, Shively G (2008) Coffee boom, coffee bust and smallholder response in Vietnam's Central Highlands. Rev Dev Econ 12(2):312–326. https://doi.org/10.1111/j.1467-9361.2007.00391.x

Dang TH et al (2001) Impacts of changes in policy and market conditions on land use, land management and livelihood among farmers in Central Highlands of Vietnam. Available at: http://sedac.ciesin.columbia.edu/openmeeting/downloads/1005791311_presentation_hapaper1forsanremconferencerevised.doc. Accessed 7 Oct 2014

De Koninck R, Déry S (1997) Agricultural expansion as a tool of population redistribution in Southeast Asia. J SE Asian Stud 28(1):1–26. https://doi.org/10.1017/S0022463400015150

Espaldon DMVO et al (2004) Challenges on Sustainable Agriculture and Natural resource management in Vietnam uplands: a case study. SEAMEO SEARCA, p 88

General Office for Population Family Planning (2010) Statistics. Available at: http://www.gopfp.gov.vn/so-lieu Accessed 12 Jan 2016

General Statistical Office (2010) Socio-economic statistical data of 63 provinces and cities. Statistical Publishing House, Hanoi

General Statistical Office (2015) Statistical yearbook of Vietnam. Statistical Publishing House, Hanoi

Hardy AD (2005) Red hills: migrants and the state in the Highlands of Vietnam. NIAS Press, Copenhagen

Hickey GC (1967) The highland people of South Vietnam: social and economic development. The Rand Corporation, Santa Monica

Hill H (2014) Is there a Southeast Asian development model? University of Freiburg, Freiburg

Human Rights Watch (2002) Repression of Montagnards: conflicts over land and religion in Vietnam's Central Highlands. Brussels. A, New York/Washington, DC/London

Institute of Sociology (2014) News of the institute. Available at: http://ios.vass.gov.vn/. Accessed 25 Dec 2015

International Coffee Organization (2015) Statistics. Available at: http://www.ico.org/about_statistics.asp. Accessed 21 Oct 2015

Kontum Province (2012) Population and labour..Available at: http://kontum.gov.vn/pages/dan-so-lao-dong.aspx. Accessed 1 Feb 2016

Le DN, Bang C (2013) Rừng Tây Nguyên đang mất (Forests in the Central Highlands is losing), Nhân dân..Available at: http://www.nhandan.com.vn/mobile/phongsu/item/20476702.html. Accessed 20 Oct 2015

Le VK, Pham QT (2014) Hướng tới phát triển bền vững Tây Nguyên (Towards the sustainable development in the Central Highlands). Trí Thức

McElwee P (2001) Coffee, christianity and conflict: an update on the central highlands of Vietnam. Mekong Update Dialogue 4(3):8

McElwee P (2008) Becoming socialist or becoming Kinh. In: Duncan CR (ed) Civilizing the margins: southeast Asian government policies for the development of minorities. NUS Press, Singapore

Michaud J (2009) Handling mountain minorities in China, Vietnam and Laos: from history to current concerns. Asian Ethn 10(1):25–49. https://doi.org/10.1080/14631360802628442

Moreda T (2012) Linking vulnerability, land and livelihoods: literature review. International Institute of Social Studies, The Hague

Nguyen TBH (2002) Phân tích thực trạng di dân đến Đăk Lăk và ảnh hưởng của nó tới sự phát triển kinh tế – xã hội (Analysing status of the migration to Dak Lak and its influence on socio-economy). Hanoi National University of Education, Hanoi

Ponte S (2002) The "Latte Revolution"? Regulation, markets and consumption in the global coffee chain. World Dev 30(7):1099–1122. https://doi.org/10.1016/S0305-750X(02)00032-3

Rev.Dr.Stan Duncan (n.d.) The International coffee crisis

Rietbergen-McCracken J, Maginnis S, Sarre A (2007) The forest landscape restoration handbook, The Earthscan forestry library. Edited by J A Sayer. Earthscan

Salemink O (2003a) Enclosing the Highlands: Socialist, Capitalist and Protestant Conversions of Vietnam's Central Highlanders

Salemink O (2003b) The ethnography of Vietnam's Central Highlanders : a historical contextualization, 1850–1990. RoutledgeCurzon, London/New York. (Anthropology of Asia series)

Sunderlin WD, Ba HT (2005) Poverty alleviation and forests in Vietnam. Center for International Forest Research (CIFOR), Bogor Barat

Thu vien phap luat (2012) Mục lục văn bản (Catalogue of documents)..Available at: https://thuvienphapluat.vn/van-ban/Xay-dung-Do-thi/Quyet-dinh-95-CP-chinh-sach-xay-dung-vung-kinh-te-moi-44156.aspx. Accessed 12 Dec 2015

Tran TQC (2007) Coffee profile. Institute of Policy and Strategy for Agriculture and Rural Development, Hanoi

UNHCR (2002) Vietnam: indigenous minority groups in the Central Highlands. UNHCR Centre for Documentation and Research

UNHCR (2006) Vietnam: situation of indigenous minority groups in the Central Highlands

United States Agency for International Development (2008) Vietnam Central Highlands needs assessment. USAID

White Paper (n.d.) The global coffee crisis

Wikipedia (2014) Dak Lak province..Available at: https://en.wikipedia.org/wiki/Đắk_Lắk_Province. Accessed 22 Feb 2016

Chapter 3
Sensitivity in Livelihood Pathways

Abstract In the vulnerability context of the indigenous inhabitants, livelihood stress is emphasized as the main pressure in this chapter. The ways indigenous people make their living are examined through a pathway lens to reveal the dynamics of their decisions. Case study, a reasonable methodology, is applied in this chapter. From October 2014 to February 2015, in-depth interviews were conducted with 60 households. The results show that although indigenous people still depend heavily on the land, land constraints do not directly affect their livelihood pathway. A holistic people-centered focus is used in this chapter to highlight the existence of household assets and the ways indigenous people mobilize these assets in their livelihood pathways. Among the assets, knowledge-experience and social position are the most important assets that affect indigenes' potential to engage in high-value livelihoods regardless of whether it is non-agricultural or agricultural.

Keywords Livelihood pathways · Land constraints · Assets · Social capital · Human capital

3.1 Introduction

The aggregative impacts of land-use change and other external factors mentioned in the previous chapter (policy reforms, population trends, and market change) on the indigenous people's life, particularly on their livelihood, call for a closer look at the transformation in their livelihood. The concept of livelihood is not new; it was used back in 1940 to describe strategies for making a living (Kaag et al. 2013). The ways indigenous people manage their livelihood, however, are not considered livelihood strategies, since their approach is not a planned means. Under the stress they face, livelihoods are rather a series of choices that emerges over time, a result of the unintended consequences of various actions. The way indigenes make their living, therefore, will be examined through the lens of a pathway. A livelihood pathway draws attention to the factors and processes that enrich people's choices when coping with shocks and constraints.

H.A.C. Thái, *Livelihood Pathways of Indigenous People in Vietnam's Central Highlands*, Advances in Asian Human-Environmental Research,
https://doi.org/10.1007/978-3-319-71171-3_3

Examining the indigenous people's livelihood through a pathways lens allows us to reveal the dynamics and different decisions. Theoretically, within the livelihood constraints, three broad clusters of livelihood strategies are identified – agricultural intensification, livelihood diversification, and migration (Scoones 1998) – and, of course, the combination of these strategies. Moreover, the scale at which an assessment takes place will be taken into account. Especially in this chapter, livelihood pathways are described at household, study site, and region levels. The different pathways between households contribute to shedding light on the dynamics of their livelihood. By contrast, the common evidence among study sites and within the region generalizes specific characteristics of the indigenous people in comparison with the other groups.

In order to gain insights into the direction of livelihood pathways and assessing factors that contribute to the choices that indigenes made, it is necessary to take a look at their livelihoods. The different ways indigenes making a living are considered as a result of their livelihood constraints due to land-use change.

3.2 Available Typology in the Households' Livelihood

Agriculture has been a major livelihood source of the indigenous people in the Central Highlands for a long time. Due to land-use transformation and accounting for constraints imposed by market, capital, indigenes have reconstructed their livelihood pattern. There are three pathways that might be generalized: farming, labor, and migration.

3.2.1 Agricultural Intensification and Diversification

Traditionally, indigenous people all engaged in agriculture, and swidden farming usually dominated their farming activities. The majority of their production was for home consumption; thus, they preferred to grow food crops, specifically rice. Nowadays, agriculture is still their dominant source of income; however, they farm using different methods.

In Kontum City, most of the indigenous farmers have stuck in traditional farming. Since neither the quality nor quantity of their land supports them to grow rice based on shifting practices, the most popular crops now are annual or industrial crops such as cassava and sugar cane, which do not require much initial investment for fertilizers or labors. In turn, however, these crops do not generate high incomes either. The imperfections in the credit system hinder them from adopting more productive, capital-intensive techniques for higher-value crops. To improve their income from these low-value crops, indigenous farmers try to intercrop yearly. Recently, a nontraditional crop named *Bời Lời,*[1] which also required very few cultivation techniques

[1] *Bời Lời* is Litsea glutinosa – a rainforest tree in the laurel family. Its powdered bark is used as an adhesive paste in incense stick production, while its trunk is used in construction. It is a perennial crop that can be harvested after 5 planting years.

and investment, has entered in their crop system. This kind of crop helps indigenous farmers improve their income; however, it may not always be as productive as cash crops in other areas. Meanwhile some villages in the Ya Chim commune own fertile land which is suitable for growing cash crops. They, however, do not have enough capital to develop it. Instead, they are forced to lease their land and then accept to be employed to work on their own land.

In a more productive land-use model, indigenes in Lak District have found stable farm work in the paddies. Thanks to the advantageous condition of fertile land located nearby the Lak Lake, indigenous farmers have been able to keep their traditional paddy crops and try to improve their land productivity by increasing the number of crops annually. However, this kind of monoculture still may not bring them a high outcome, since the price of rice is increasingly dropping. Normally, the outcome of every harvest is just enough for farmers to cover their investment costs and their own food demands until the next harvests. Sometimes, the rice outcome may not be enough to purchase essential goods and services, which are increasingly commoditized, or to pay for higher needs such as school fees, medical cost, etc (Fig. 3.1).

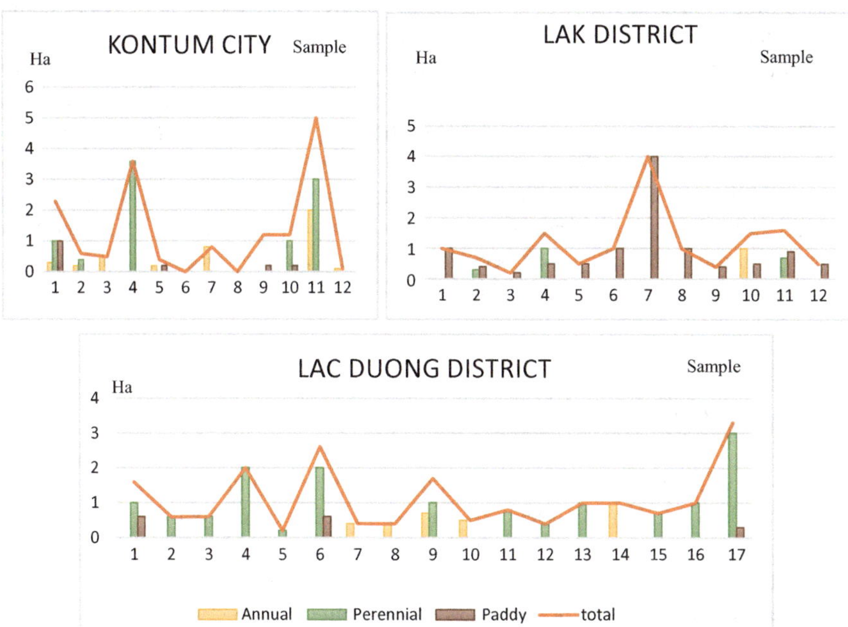

Fig. 3.1 Sharing of land into diversified crops of samples at Kontum City, Lak District, and Lac Duong District (*Source*: Interview at Kontum City, Lak District, Lac Duong District from Oct–Dec. 2014, drawn by author)

In order to escape monoculture and improve their income, indigenes are encouraged to improve the productivity of land and diversifying into commercial crops. This pathway, however, also involves land restriction. Since Lak District has received the most migrants among three study sites, the conflicts over land is inevitably the most serious here. In fact, the development of cash crops followed the rise of migrations in this district. By the time the local people recognized the value of cash crops, in particular coffee, the most suitable land had been already occupied by migrants who have claimed fallow land or bought land cheaply a long time ago. Ironically, some indigenous people in Lak had to buy land back from migrants. However, for most indigenes, the desire to own cash crops is thwarted by their lack of capital to purchase land and invest in initial crops. Therefore, most of the cash crops in the Lak District belong to migrants. Besides farming, the indigenous people there have traditionally raised livestock as an important source of income, but the current scarcity of grassland also makes it difficult to graze cattle.

In Lac Duong District, most indigenes own their perennial and annual crops. The subsistence or traditional farming seem to have been entirely swept out, as indigenous farmers have shifted to more intensive crops such as vegetable, fruit, and coffee. However, the land distribution disparity with the Kinh people unfortunately still exists. Thanks to the fertile soil and the comparatively colder climate, this district is suitable for developing high-tech crops such as flowers; however, most of the flower areas in this district are owned by the Kinh people. The lack of knowledge about farming techniques and capital hinders indigenous farmer from learning from their Kinh counterparts. For the indigenous people in Lac Duong District, the need for land is not always as high as the need for capital. The concern regarding land use in Lac Duong District is not exactly related to the amount of land the indigenous people have access to but rather to the way they use the land. Many households own their land but cannot use it productively. Similar to cases in Kontum City, indigenes in Lac Duong District rent their good-quality land to Kinh farmers who would set up flower gardens, and they then become day-salary earners and/or cultivate coffee on the remote, sloping, less-fertile land.

3.2.2 Labor-Oriented Income

Contrary to previous images of indigenous household as pure farmers, the indigenous people now, due to the asset constraints on farming, try to perform many activities to relax credit and liquidity constraints. As a common trend of rural households, highlanders are participating in labor markets, in self-employment or wage employment in agriculture or in the non-agriculture economy (The World Bank 2008, p.75). These different activities are discussed below.

Handicraft Production

Like other ethnic minority groups in Vietnam, indigenes in the Central Highlands have some specific handicrafts. Brocade weaving and basket weaving are two that have been maintained so far. In the past, the majority of their production is for home consumption. Nowadays, these have become commodities from which indigenes could earn an extra, albeit not a stable one.

Among study sites, Lac Duong performs the most successfully. The communities in Lac Duong have developed their traditional handicraft products into a market product. The authorities allow them to do business with their weaving products at tourism destinations. Some households even live on their handicrafts that do not happen at the other study sites. However, in this market-oriented economy, handicrafts face many challenges: the competition with industrial products and the lack of natural materials. Indigenes sometimes mix industrial products with handicrafts at their market stalls. Authorities at the study sites have organized some programs to encourage indigenes to conserve and transfer this traditional craft to the young generations. However, not many young people are interested in maintaining their handicrafts, especially as a livelihood.

Waged Labor

Besides the major occupation in farming and the small income generated from the traditional handicraft, some indigenous households manage their livelihood constraints by wage employment. The wage workers' situations are varied, either related to agriculture or non-agriculture.

The labor market is the most abundant in Lac Duong District, where agriculture is the most productive. Indigenes are hired to work on farms, most often to harvest strawberries. They earn 150.000–200.000 VND (US$ 6.65–8.87) (US$ 1 = 22.546 VND in 2015) per day. Their wage is paid after every working day, and waged workers do not have any guarantee or contract. However, most are satisfied with this work. Indigenous people at Lac Duong have their own coffee gardens; thanks to the good climate conditions, farmers do not need to spend much of their time on watering or looking after their coffee gardens. Spending free time on agricultural wage brings them an extra income; some indigenous households in Ya Chim commune, Kontum City, also specialize in wage employment. They work on their land, which is being rented by private or state companies to invest in growing rubber. *A renting period can last 10 years. During this period, the indigenes could work and make a profit in the ratio of 4:6 (the indigenes receive 40% profit from rubber; the rest goes to the company who made the initial investment)* (interview at Plei Bur Village, Ya Chim commune, on 29 December 2014). In these two cases, employment is relatively stable, as these jobs are offered yearly, and indigenous farmers can work with the available skill they have, and at least they own their asset: land, which is only rented out temporarily (Fig. 3.2).

Source: Anh Chi Thái Huỳnh *Source:* (The World Bank 2009)

Fig. 3.2 Handicraft products of indigenes in the Central Highlands (*Source*: Huỳnh Anh Chi Thái)

Meanwhile, at other areas in Kontum City and in Lak District, where the labor market is restricted, indigenous groups are forced to apply for low-value nonfarm employments, because they have not enough land and have insufficient seasonal farming income to survive in the off-season. In this case, the original indigenous farmers try to get any jobs they are offered; most of these jobs earn them an unskilled wage and involve hard work at construction sites and road buildings. This kind of jobs is not constant and may not generate a secure yearly or even monthly income.

Public Sector Employment

Naturally, people who are educated have more chances to find a stable job. Recognizing how hard farm working is and how restricted access to land can be, indigenous people have come to increasingly appreciate the value of education and are therefore investing more in their children's education with the hope of this helping them overcome poverty. In particularly in Lak and Lac Duong District, many young people are able to go to university. However, another concern that has been raised is the lack of transparency in recruitment processes after young people have graduated. This is a common problem in Vietnam and likely more serious in these poor, less developed study sites where private sector jobs are rare. The state sectors are the dominant employers here. The most positions that indigenous children can apply for are jobs in teaching, medical staff, or other staff jobs at local state offices. Indigenous households are deeply disappointed, as many of them have gone into debt for their children's education, and have sometimes ended up having to pay more to get a job. The young generations feel discourage when they witness how many graduated villagers cannot get jobs, and this also makes the parents hesitant to invest in their children's education.

Mr. Them, Bon Dong, Lac Duong District

We are miserable enough, hardworking enough, do enough farming. I hope young generations in my village could go to school to better their future life. However, the youth do not like it. They feel discourage. No matter how well they study, it is difficult for them to get jobs. All jobs are given to descendants of influential families. People claim 100 million VND for a position. It is unfair, as we have to go into debt for our children's study, and now have to pay even more to "buy" a job. If we had this of money, we would invest it in our business rather than use it to pay to be hired at their offices.

Services Sectors

Not as many indigenes work in the service sectors as in the trade or tourism sectors. Even at the village level, petty trading is dominated by the Kinh people. There are, in fact, some households who have engaged in trading. Their business is relatively various: small shops, selling second-hand motor bikes, hair salon, etc. However, not many of them achieve success in business. Most indigenous business does not last long and is replaced by the Kinh majority. According our observations at three study sites, there is only one shop at Le Village, Lak District run by an indigenous household.

Likewise, in the tourism sector, as CBT is a kind of tourism, which is based on community, and for their benefit, indigenous communities are encouraged to join. However, they seem not to choose tourism at their own initiative. Indigenes are unlikely to join tourism activities unless they are triggered by outside forces. Until now indigenes at all three study sites have joined tourism development projects; however, not so many have tried to run tourism programs by themselves. The role and participation level of indigenes in tourism activities will be discussed further in the next section.

3.2.3 Migration-Oriented Income

Whereas migration is recognized as a popular choice for farmers living under livelihood constraints (Hussein and Nelson 1999), it seldom occurs in the Central Highlands. Among the study sites, migration is mentioned as a minor option for indigenes' livelihood. Any migration that does occur is predominantly among young people. Most of them are unskilled laborers; they may find *simple jobs with a low salary in spite of migrating to big cities like Saigon* (interview at M'lieng 2 Village, Lak District, on 10 December 2015). Given the high cost of living in the city, the fund indigenous migrants are able to send back home is not remarkable. Normally, they *spend a short time of their youth in wage labor in big cities; then, they come back to get married and settle again in their villages* (interview at Nam Pa Village, Lak District, on 13 December 2015).

The low salary, hard work, and, more importantly, the racial discrimination that they encounter during their time as migrants are some mentioned as reasons that drive them to return home. The main migration trend to big cities in Vietnam right now is of Kinh people who are from rural areas or poor provinces. Migration is not a stable solution for indigenes.

The livelihood pathways in Vietnam's Central Highlands are much different from the common framework of rural livelihood (Chambers and Conway 1991; Scoones 1998; Hussein and Nelson 1999). The framework that simulates routes to sustainable livelihoods pursued by rural people in situations where there are constraints caused by ecological or socioeconomic conditions is normally established by these elements: external conditions, livelihood resources, livelihood strategies, and livelihood outcome (Scoones 1998). So far we have discussed external conditions (in Chap. 2), which included policy reforms, population trends, and market changes. The livelihood portfolios mentioned in this chapter reveal some of the following common points. Like other vulnerable case, livelihood strategies of indigenous people adequately encompass three broad clusters: agricultural intensification/extensification, livelihood diversification, and migration (ibid). Their livelihoods generally still heavily base on land in spite of the constraints of this resource. According to observations at the study sites and the information from previous research (United States Agency for International Development 2008; The World Bank 2012), the livelihood outcome of indigenes is judged to still lag behind in comparison with other groups in the same area. Based on all above, the framework of a rural livelihood can be adapted into case of the Central Highlands as Fig. 3.3.

In this framework, the remaining elements not yet analyzed are institutional and organization structures. This element differentiates livelihood framework in this case study from the common livelihood framework and strongly decides livelihood outcomes. Institutional and organization structures are determined by historical context and cultural characteristics and will be detailed in the following section.

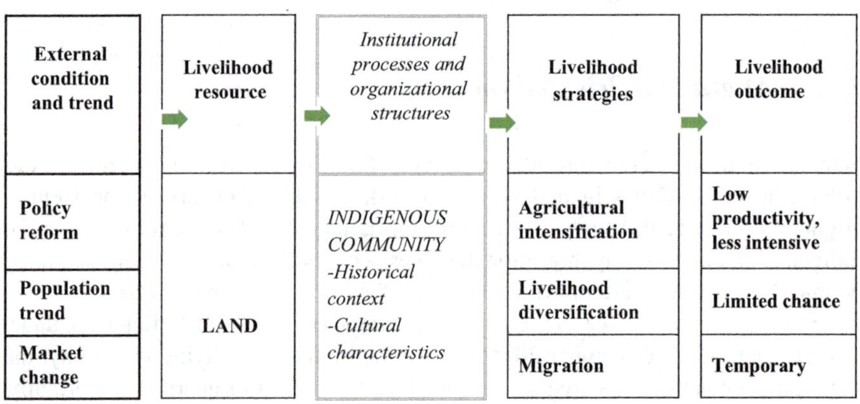

Fig. 3.3 Livelihood dynamics: a framework for analysis (Adapted from Scoones 1998)

3.3 Factors Forming Livelihood Pathways: A Region Scale Approach

Livelihood in this study is investigated under a vulnerability context of land constraint. Vulnerability and livelihood are viewed as two sides of the same coin (van Dillen 2004). In the previous chapter, we analyzed the external side of vulnerability; in this section, we will view livelihood more from the internal side. Although the statement "rural livelihood strategies are often heavily reliant on the natural resource base" (Scoones 1998, p.11) is definitely true for this study case, it cannot be denied that a livelihood portfolio is a choice based on people's characteristics, and land constraints are not the only driving force. This argument is supported by causal structure of vulnerability, in which Bohle states that "vulnerability is a multilayered and multidimensional social space defined by the determinate political, economic and institutional capabilities of people in specific places at specific times" (Watts and Bohle 1993, p.4). This can be clarified further when making a comparison between the high agricultural outcomes of the Kinh majority and the less productive outcomes of indigenes, though they share the same natural resources. Even the latter have more advantage in some cases. In other words, the internal side is identified mainly as an orientation to the livelihood pathway and will shed light on and explain the livelihood pathway differential.

Also, the way that indigenes set up their livelihood pathway and are lagged behind the other groups is viewed through a sensitivity aspect – one of three phases of the vulnerability concept (exposure, sensitivity, and resilience) (Turner et al. 2003). This concept is nothing more than the internal capability/characteristics (Prowse 2003) when reacting to the external phenomena mentioned in the previous chapter. The sensitivity approach is deployed here to emphasize "the risk of being susceptible to stressors" (Warner 2006) or being unable to cope with adverse effects (Brooks 2003). Linking sensitivity to the livelihood pathway can enhance our understanding of the limitations of villagers due to their livelihood characteristics, of how different indigenous households make their living, and of how they may not be able to generate good outcomes in comparison to other groups.

The internal side or sensitivity aspect, which derives from historical community context and specific cultural characteristics (Füssel 2007), will contribute the core of transforming and processes that influence livelihood strategies and constitute the last element (institutional processes and organizational structures) to complete the livelihood pathway framework as mentioned above. Even though some of the evidence mentioned here might be anecdotal, it does shed light on the indigenes' situation.

3.3.1 Structural Approach: Historical, Economic, and Cultural Characteristics

We do not intend to explain the disparities in livelihood between indigenous people and other groups based on their cultural or historical characteristics. This kind of differences between indigenous people and other groups is rather simple, since these groups are not homogeneous. However, the involvement of institutional actors largely relates to the power, and processes, and is closely linked to long-term social, political, economic, and cultural factors (IFAD 2009). Furthermore, as external observers have criticized the Vietnamese government and donor agencies for pursuing assimilationist policies and ignoring cultural identity among ethnic minority groups (Valerie 2014, p.177), this study purposively takes a historic and cultural approach to reach the first factor and set up it as a background for the more detailed actor-oriented factors, which will be laid out and discussed next. In other words, we will analyze the structural side and subsequently its opposite side: the actor aspect.

Sensitivity from Traditional Economic System's Characteristics

The specific choice and shifting livelihood pattern of indigenes has a cultural and economic aspect (Valerie 2014). Agriculture was virtually their sole means of livelihood, and a self-sufficient system was predominantly practiced. Ethnic groups in Southeast Asian may be divided roughly into two major segments: "the people of the plains, valleys, and deltas, who have been strongly influenced by the civilizations of India and China, and the inhabitants of the highlands who remained far more aloof from the great current of history" (Hickey 1964, p.v). Thereby, whereas the lowlanders had a rather modern economic system, the highlanders in the Central of Vietnam still practiced the economic system of exchange, also called the barter economy. With a rather isolated society, this system helped indigenes to exchange goods and meet their needs, including agricultural products, handicrafts, and some higher-value property such as land, clothes, jars, buffalo, and even labor, especially during the harvest seasons. For centuries, indigenes led a distinct economy in comparison to the Kinh majority's system.

This pattern is rooted deeply in their society and still spreads its impacts in their modern society (Hickey 1964, p.vi). Even until the 1970s, money was not a dominant factor in all indigenous communities. The impact of this system still echoes in the more recent modern economic system. For instance, poor households who could not afford to purchase cows may be hired to feed cows for richer households. *Hirelings are not paid directly; instead, when the cows give birth, people can own one baby cow as their wage* (interview at Nam Pa Village, Lak District on 13 December 2014). The mutual labor can be arranged in farm working among relatives or villagers, including the peeling of cassava, drying and cutting cassava, harvesting coffee, and assisting during parties and funerals. This reciprocal relationship is increasingly disappearing in terms of frequency, trust, and exchange networks and is now practiced among relatives only. However, the persistent existence of the barter economy in highlanders' society obviously leaves negative impacts on their economic activities, as the market economy currently prevails, and calls for a considerable change in the pattern of the indigenous economy. Indigenes are required to master new farming techniques, marketing methods, organization of transport, small business techniques, and ways to invest effectively in crops. With a lack of credit and knowledge to manage their finances, households struggle to expand agricultural productions or diversify their livelihoods.

Most of indigenes lack credit to invest in their crops and have difficulty accessing to credit. They generally prefer to lend from the informal sector than from state banks. This informal lending is available in every village, is simple to access, and requires less paperwork, whereas the cumbersome procedure of bank lending likely confuses indigenes and is time-consuming. Normally, the loans indigenes take out in this case are small and are enough to cover the initial investment for crops and sometimes for their daily needs. This preference, however, pushes them into risk. Because they become increasingly dependent on private money lending, and are forced to pay high interest after every harvest. Their lack of language skills and lower experience levels with calculation make them especially vulnerable when dealing with money lenders (The World Bank 2009).

Also, indigenes are not familiar with managing large amounts of money; thus, it is understandable that they panic, as they get into bad debt. In this case, indigenes are more unlikely to make wise decisions and instead end up in serious debt situations. This also explains why many highlanders have sold their land to repay their debts, in particular during the coffee crisis. Land is their only unique property which can easily and quickly be exchanged for cash.

Mr. A Lu, Kon Ro Bang Village, Kontum City

We – the ethnic minorities – are not as wise as the Kinh majority. Our ancient land was large, then it was divided into small plots for new generations. In particular, during the subsidizing period, we were so poor and hungry, so we sold our land to buy rice. We are still not good at calculating. We sell our land to buy motorbikes, to build houses, then we become wage-earners.

Another troubling situation is when indigenes possess large amounts of money, mostly from selling land or receiving clearance compensation from construction projects. The money seems to come easily in these cases, but unfortunately this capital is not invested optimally. Normally, the money is quickly wasted on short-term and minor needs. Many of indigenes, for example, *sell their land just for their child's wedding* (interview at Kon Jori Village, Kontum City, on 22 December 2015). Some other typical cases can be pointed out at Kon Ro Bang Village, Kontum City. Located nearby the city center, this indigenous community is strongly influenced by the Kinh majority. The young generations want the latest fashion and new motorbikes, whereas the older wish to build Kinh-style houses. They sell their land to make these purchases and subsequently try to manage their life with only a restricted plot of land and a lack of capital.

Three different cases above portray indigenes' weakness in using capital. This sheds light on the opposite pathways the indigenes and migrants follow for their livelihood. Early observations at study sites have aroused concern and raised the question: *Why migrants who have only recently resettled in the uplands with very little knowledge about the physical conditions of this area could set up a successful farm with a much higher outcome than indigenes*? Most indigenous interviewees explained that it was their lack of capital why they ignored high-tech crops, though these crops promised to bring high outcomes like they did for the Kinh. However, this is not a sound reason, since many high-tech crop producers were poor when they started out and did not even own land.

As Swift (1989) states: investment occurs when production leads to a surplus beyond consumption requirement. Indigenes were unfamiliar with the concept of saving money. They have still not gotten used to long-term investment, which requires good planning, careful calculation, and long period of waiting until profits are earned. Also, very few have even handled a large loan. As a result, indigenes are unlikely to dare to invest a lot of money in a venture such as high-tech crops. Additionally, the indigenes dread having the burden of a big debt due to investment in intensive crops. Meanwhile, the important strength of migrants is that they have experience of producing intensive market crops in their home lowlands. This means that lowland farmers are more willing to take the risk of getting loans and set up their new life in the uplands.

In this market-oriented economy, a lack of access to capital and inability to set up a good venture are real constraints that hinder indigenous farmers from restructuring their agriculture system. Normally, they prefer to persist in their traditional crops than to transform to new ones or diversify their livelihood. The situation of lacking skills in financial management is even worse, as it is added to their restricted access to markets. Since the role of markets is extremely important in a market-oriented economy, the unfamiliarity of market function places indigenes in even more vulnerability, and they become more likely to be taken advantage of.

Sensitivity to Cultural Characteristics

Another side is that the livelihood pathway feature is partly characterized by the indigenes' specific cultural background. The difference between indigenes and the Kinh majority is a subtle problem and not easy to measure (The World Bank 2009).

One of the first factors is based on misconceptions people hold about indigenes. According to the Vietnamese government, the ethnic groups are backward thinking, uneducated, lacking aptitude for business and market interaction, innately inferior, and with old cultivation habits (Hickey 1967; McElwee 2008a; UNHCR 2006). They believe that the advanced lowlanders can help indigenous people improve their lives. These stereotypes were already around in colonial times, but it seems even more highlighted since the Vietnam War and "serves to disempower or deprive minorities of their ability to full participation in society" (The World Bank 2009, p.43). Indigenous people sometimes lose their self-esteem, self-confidence, because of this. They themselves also start to believe that they are backward and worth less than the Kinh majority. One interviewee stated that: *The Kinh people are smart and quick-witted; we are inferior* (interview at Kon Ro Bang Village, Kontum City, on 26 December 2014). Another remarked that: *The Kinh people are superior, since they own both knowledge and capital. Meanwhile, we are poor and know little. It is so miserable* (interview in Kon K'tu Village, Kontum City, in 22 December 2014). Stereotypes held by the government and the lost self-esteem of communities contribute to discourses of the Vietnamese government becoming self-fulfilling prophecy. However, the situation of the indigenes cannot be explained based only on their cultural characteristics but requires us to look at a variety of aspects.

Another outstanding aspect of common patterns of the highland people is the fact that their societies are village-centered with a high level of intra-village cooperation (Hickey 1964; Valerie 2014). The village has a strong collective spirit. And the close-knit communities more or less influence their livelihood pathways. As mentioned above, their strong community relationships mean they prefer to stay comfortably in their village, rather than emigrate elsewhere for work. Even some who have already migrated plan to return soon. Indigenes find most dignity and enjoyment in life when living within their communities. One interviewee bitterly described this as follows: *In the big cities we look just like monkeys who have moved away from the forests and are lost on the streets* (interview in Kontum City, Lak District, in December 2014). The very few economic benefits are unlikely to compensate for the social cost they pay when moving away from their home community.

Besides, social relations in indigenous community are closely involved in and control their transactions. In particular, the idea of community reciprocity creates barriers for indigenes to engage in and earn a profit from trading (The World Bank 2009, p.41). There are very few shops in minority villages owned by indigenes, and normally the shops of indigenes are unable to operate for long. When I asked villagers, including some who have opened a shop before, about this circumstance, again and again I was told that they cannot successfully run their shop because of cultural barriers. Indigenous owners are besieged with requests from kin and co-villagers for purchasing on credit (McElwee 2008b). *When they do not have enough money, they come to my shops, take stuff, and leave behind a debt. It is difficult to get them to repay the debt, but I cannot refuse them. Co-villagers should not displease each other. But when they have money, they prefer to go trading at the Kinh shops* (interview at Le Village, Lak District, on 01 December 2014). Another problem is that whereas the Kinh majority described indigenes as "not having a head for business" (The World Bank 2009, p.41), indigenes themselves also admitted that they were not used to and could not quickly adapt to the numerous tricks that are used in the marketplace. Such tricks are necessary to make a profit and be successful in a market economy, even though they go against the indigenes' moral code and thus cannot be applied within the communities.

Last but not least, regarding cultural characteristics, it is assumed that the longstanding free and liberal life that indigenous people lead in the midst of the forest with a range of festivals is another barrier for them to internalize modern life and become a part of the market economy. People believe that this cultural backdrop is also the cause of a working style that involves the less discipline. Highlanders, especially the older generation, used to lead a less competitive life, and they are not familiar with the modern working environment. They feel restricted when having to work according to strict regulations under the control of managers. During a field trip, we met many highlanders who preferred the hard life as a farmer than finding a job. They explained that *we love our freedom, and we would like to earn our own income. If we have to work for someone else* (*làm thuê*), *it means we will be forced to follow their rules. We are forced to come and leave on time, work according to regulations. We do not like it* (interview in Da Blah Village, Lac Duong District, in 18 December 2014). That is why many have ignored opportunities to work at the factories or for companies. For them a free and easy-going life is sometimes more important than wealth.

In return, this characteristic makes employers hesitate to employ highlanders. A priest of an indigenous parish in Kontum City explains the situation of his parishioners as follow:

> *Many young people in this village are unemployed. In the past, they owned much farmland, but that has changed. Now even food security – they are not able to guarantee. Apart from farming, not many of them have a stable job. There is a sugar cane factory in this village, and villagers are sometimes employed as the security. Nevertheless, the managers complain about their lack of discipline so often. Today they work, tomorrow they off. So arbitrary!* (Interview at Kon Ro Bang Village, Kontum City, on 25.12.2015).

The common choice and similar circumstance of indigenous livelihood pathways unveil a sensitivity derived from their socioeconomic and cultural patterns. This affirms that the physical or biological hazards represent one pressure, and a further pressure comes from root causes of social differentiation and cumulative progression (Wisner et al. 2003, p.5). After around 40 years, the Vietnamese government has implemented policies of settlement and sedentarization with the hope that "Kinh and indigenous would live together in ethnic solidarity" (McElwee 2008a, b. p83). It is unfortunate that there still exists such a great difference between these different ethnicities. However, it is superficial not to consider other aspects to explain their job decisions, particularly that of younger people, and the underpinning factors relating to social discrepancy, and their less economic benefit that really discourages indigenous people from making their living effectively should be added and will be discussed next.

3.3.2 Delineation Beyond Ethnic Boundaries

The disparities between indigenous people and other groups, in particular the Kinh majority, derive from more than just historical characteristics. In this section, we will address the context and driving forces of social life that separate indigenes from other groups and influence their livelihood decisions.

Indigenous People and Migrants

Disparities in Access to Education

One major rationale to explain the gap between indigenous groups and others is their generally lower levels of education (UNHCR 2006). For a long time, indigenes have had the highest rates of illiteracy in Vietnam (Human Rights Watch 2002). Besides economic policies implemented in this area, the Vietnamese state has identified education as a prior target in the Central Highlands, which is expected to lessen the gap and inequalities between indigenous group and the others. The authorities have showed their efforts through expanding the school system in these upland areas, encouraging indigenous pupils to attend by exempting them from school fees, giving them books and pens, initiating scoring incentives, posing lower entry requirements, and implementing quotas for minorities (Human Rights Watch 2002; Valerie 2014).

The state programs for enhancing education levels have had good results. Illiteracy rates have been steadily decreasing (Dang 2010). The Vietnam Household Living Standard Survey (VHLSS) of 2010 shows that primary school attendance of indigenous children in the Central Highlands is good, even better than in some other areas. However, these figures only reflect the rates, not the quality of education children receive. In reality, *5th grade, even 7th grade students cannot read fluently in Vietnamese, yet they are allowed to progress to the next level* (interview at Dung K'No Village, Lac Duong District, on 17 December 2014). According to stories from interviewees, we know that this is not a single case. Even more unfortunate is that the re-illiteracy sometimes occurs, as the children drop out after finishing primary school. This arouses skepticism about official results.

In reality, the relative gap of education levels between indigenes and other groups still exists. Inequality in access to schools is one explanation for this. Indigenous communities are often located in remote areas. Nowadays, primary schools have been built in each district in these uplands to make sure that indigenous children are able to make daily trips to school, so the dropout rates at primary school are low (Valerie 2014). However, at upper secondary level, the pupils have to leave their family and move to a district town to go to school. This brings about a lot of preparation and a great many challenges: renting room, bringing rice and food from home, and other daily expenses. The long distance between upper-level school facilities and villages strongly influences the attendance of indigenous students. "The remoteness of their villages discourages them from persisting in school" (United States Agency for International Development 2008, p.15). This is reflected in the attendance reduction of indigenous children in upper schools.

Beside the long distance to school and excessive school fees that cause the dropout rate to increase of indigenous students, the low benefit young indigenes experience from their advanced schooling does not motivate them to persist in school either. As far as we have seen, indigenous students find it difficult to find a wage job after graduating, and they sometimes even have to *pay to buy a job* (interview at Bon Dong Village, Lac Duong District, on 09 December 2014). They often have no other option than to return back the fields (although agricultural land was the "push" factor that led them away from farm and toward an education target), and academic knowledge is not necessary for their farm work at all. This generates a vicious circle of indigenous disadvantage (Table 3.1).

Economic Motivation of Newcomers

"Vietnam was one of the poorest countries in the world, [and] economic growth was stagnant" after the Vietnam War (Dang et al. 2003, p.4). This situation stimulated people's mobility, and they went in search of new economic chances. A massive influx into the Central Highlands was the main direction. The immigration to the Central Highlands reflects one of the major movement patterns in Vietnam at that time: rural area to rural area.

As it has been discussed in Chap. 2, there were two immigration trends into the Central Highlands: the government-organized resettlement to NEZs and the spontaneous movement. However, no matter which trend they followed, most migrants originated from the same status. The migrating direction to the Central Highlands during this period was chiefly from Northern Vietnam, where the population density is high, there is a lack of agricultural land and land is often unsuitable for cultivation, and natural disasters often occurs. Therefore, from the very beginning of their movement, migrants have aimed to start a better life and exploited new areas of land. Migration is a big compromise for many people, as they have to leave their relatives. The Central Highlands at that time was known as an "untouched area" (McElwee 2008a, b), with very poor transport connections and communication facilities, making migrants even more separated from their hometowns. Furthermore, many of them had to *sell their land to get capital to settle down in a new location*

Table 3.1 Net school enrollment rates in Vietnam in 2010

	Primary			Lower secondary			Upper secondary		
Region	Male	Female	Total	Male	Female	Total	Male	Female	Total
Red River Delta	95.0	93.5	94.3	89.6	91.9	90.6	69.2	67.2	68.2
Northeastern mountain	93.0	90.9	91.9	85.2	83.0	84.1	56.0	60.7	58.3
Northwestern mountain	93.3	93.9	93.6	80.9	65.5	74.2	47.4	38.8	42.7
North-central coast	90.0	91.1	91.0	83.8	87.6	85.8	54.7	58.9	56.8
South-central coast	92.1	90.7	91.4	89.5	86.4	88.1	58.4	69.6	64.0
Central Highlands	**95.4**	**87.7**	**91.9**	**67.3**	**78.2**	**73.1**	**45.6**	**52.5**	**49.1**
Southeast	90.3	97.9	94.1	76.1	81.8	78.4	52.8	63.1	58.0
Mekong Delta	91.4	92.7	92.0	66.1	76.5	71.2	39.2	50.5	44.1
National	92.5	92.8	92.6	80.0	83.3	81.5	53.9	60.1	57.0

Source: Vietnam Household Living Standard Survey

(interview at M'Lieng Village, Lak District, on 08 December 2014). Beginning again with empty hands placed even more stress on them to succeed (Table 3.2).

The pull and push factors are even clearer for spontaneous migrants than the previous migrant groups. Economic motivation has more clearly been identified as the main reason of their movement. The high outcome of cash crops in the Central Highlands in the early 1990s spurred more new migrants, who had careful plans when moving to these uplands. Many of them were the relatives or neighbors of the previous groups who had settled in the Central Highlands; therefore, they likely learned from the experience, skills, and techniques from each other to apply quickly on their farm. Even if they had never grown cash crops before, migrants had access to many channels to learn and apply their new knowledge and skills better than the indigenous people. There were many networks supporting migrants in their access to information, finances, and power, and these were the main keys that led to the high economic outcomes among Kinh migrants (The World Bank 2009, p.28).

Intensive Agricultural Skills of Migrants from High-Density Population Areas

Compared to how the migrant transformed from intensive annual crops to cash crops, the indigenes had far more difficulty and took much more time transforming from subsistent crops to commercial crops. Most migrants were from high-density population areas, and they were used to developing intensive crops (two or three crops per year, using fertilizers and other chemical substances). This background helped them easily adapt to the new cash crops in the Central Highlands. When coping with crop failures, in particular during coffee crisis, the migrants also had more experiences to manage these events cleverly. In one research studying on how farmers coped with declining coffee prices, researchers found that "Kinh migrants are more flexible, they then are likely to change crops, liquidity enhancements, reductions in inputs, or

Table 3.2 Out and net migration rates of regions in Vietnam 1999–2009

	Migration-in rate[a]		Migration-out rate[b]		Net migration rate[c]	
Region	1999	2009	1999	2009	1999	2009
Northeast	16.15	15.9	27.53	33.5	−11.38	−17.5
Northwest	13.24		14.57		−1.32	
Read River Delta	23.28	35.0	32.61	36.7	−9.33	−1.7
North-central coast	8.61	16.0	31.97	50.6	−23.36	−34.6
South-central coast	17.02		29.74		−12.71	
Central Highlands	**86.24**	**43.3**	**16.22**	**32.1**	**70.02**	**11.2**
Southeast	68.33	135.4	26.80	27.7	41.53	107.7
Mekong River Delta	14.71	16.3	24.59	56.7	−9.88	−40.4

Source: United Nations (2010)
[a]Migration-in-rate is a ratio of number of migrants-in over total local inhabitants (per thousand)
[b]Migration-out rate is a ratio of number of migrants-out over total local inhabitants (per thousand)
[c]*Net migration is a ratio of net migrants (migrants-in minus migrants-out) over total local inhabitants (per thousand)*

engage in off-farm employment" (Dang and Shively 2008, p.314). None of the migrant households reported selling land (ibid: 319). Meanwhile, when the first commercial crops of indigenes failed, they were unable to respond appropriately, and many lost their properties to pay for debts and lost their courage to begin again.

Influences from Outside Forces

Physical Conditions

At a regional scale, the historical context and distinct cultural characteristics of indigenous communities have been analyzed as the factors that determined similar types of livelihood pathways. Besides that, the context of choosing livelihood pathways is also embedded in outside forces. Although these forces are not embedded in the internal sphere of indigenes, they are discussed here to explore the livelihood pathways on a regional scale.

We can distinguish different livelihood pathways as derived from the different physical conditions among the study sites. In Lac Duong District, for example, there is fruitfulness and wealth due to the high productivity crops. Skills and favorable social context alone do not explain this; but the natural advantages in this area do shed more light on the matter. The fertile soil and suitable cold climate have helped the high productivity of intensive crops in this area. Meanwhile, the poorer natural conditions in Kontum City have created challenges for intensive agriculture.

Location and Accessible Ability

An equally important issue is the location of the different study sites. In this market-oriented economy, an accessible location plays an important role. Being closer to a market obviously promotes production and provides people with access to more information so they can determine which products are most appropriate for the market. Lac Duong District again has this advantage, as it is located near big markets like Ho Chi Minh City. In contrast, the rather isolate locations of some communities in Kontum City, and Lak District, prevent local farmers from accessing markets and create a lack of value-chain links between producers, traders, and consumers.

In general, the Central Highlands is often portrayed as a remote or even untouched area (McElwee 2008a, b), and the poor infrastructure strongly influences their chance of development. "Poor infrastructure leads to poor education, poor employment, and then poor access to markets and services" (Valerie 2014, p.175). And the disparities in access to infrastructure are quite clear in these uplands. Up to now, the different groups have different living arrangement with indigenous communities remaining in their distinct villages (ibid).

The marginalized situation of indigenes can be explained by the following causes. Traditionally, highlanders were used to living on the uplands and practicing shifting cultivation. During the resettlement and later the urbanizing process, they had the chance to move to lower land or relocate closer to the main roads. However, they *preferred to keep their ancestral land and live on it* (interview at Le Village, Lak District, on 01 December 2014). Meanwhile, most of the migrants from rural lowland areas were attracted to these more accessible areas, as they were unfamiliar with uplands, and were reluctant to move to the isolated forest areas. In addition, when the newcomers wanted to purchase land, the indigenes were *willing to offer their conveniently located land to get a higher price* (interview at Da Blah Village, Lac Duong District on 18 November 2014), and they then accepted to cultivate in more remote areas. In other cases, when a newcomer (usually the Kinh majority) moved into an indigenous village, they would often aim to take on role such as shop owner. In order to do so, newcomers chose to settle in the most suitable locations: at the beginning of each village (village gate). As well, the towns in each district usually have a high percentage of Kinh people. In my observation, this arrangement of living space is similar in the different areas: the indigenes tend to live in more distant areas, while newcomers settle at more favorable locations.

In this way, indigenes marginalize themselves economically and distance themselves from social services (United States Agency for International Development 2008). Indigenes still have relatively limited access to public utilities, in particular, water and sanitation facilities and electricity, for example. This disadvantages them in economic term, because the inconvenient transport conditions and poor access to information keep the community isolated from the market and prevent them from gaining direct access to transactions between demand and supply sites. The poor connecting transport facilities also partially attribute to their lack of mobility. Highlanders have become used to limiting their living space in their local area, which means they have less chances to access the wider world and get exposure to information and adapt to new developments Fig. 3.4.

In other cases, good locations closer to markets and roads have meant the indigenous people's lives seem to be less peaceful because their communities are easier to be touched by outsiders and have to cope with severe competition from the Kinh majority. In this circumstance, communities can learn from the Kinh people to intensify and diversify their livelihood. However, this also creates extra pressures for them and sometimes leads to negative influences. One particular case is the Kon Ro Bang Village, located next to the center of Kontum City. Villagers here learn more quickly from the Kinh people than the other groups. Their lives have become more active and modern, and their livelihood is more diversified, as they have been able to better develop their service and trade. However, the young people here are attracted to gambling and other entertainment services. Villagers have even sold land so they could access this entertainment or invest in developing services of their own. This leads to them being forced to live in cramped conditions and trying to make ends on a tiny plot of land. The villagers produce neither fish nor fowl, as they are not exactly urban residents, and they do not do well in business or trade and lack land for farming.

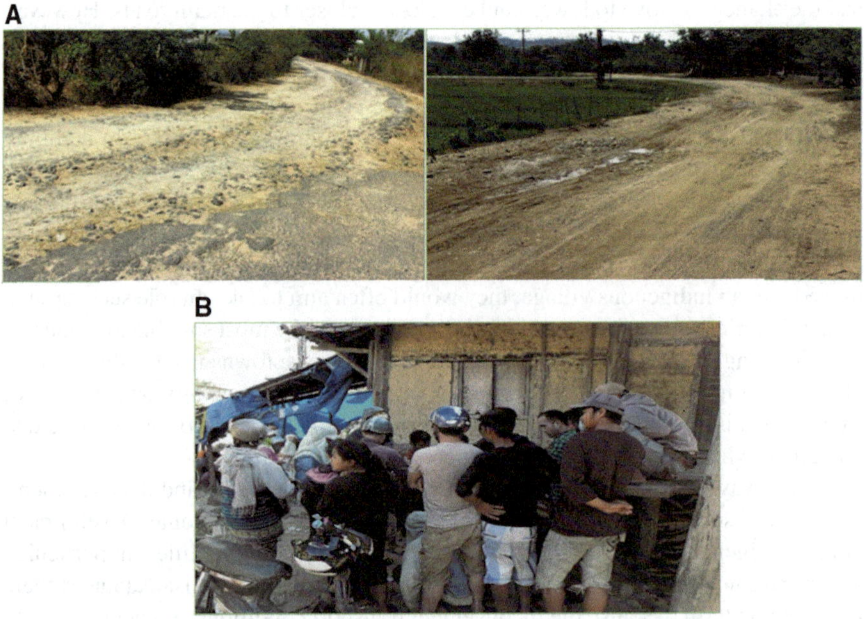

Fig. 3.4 Location and its influence on indigenous peoples' lives (**a**) Poor infrastructure to access Kon K'tu Village, Kontum City (*Source*: Author), (**b**) Gambling at a well-accessible village (*Source*: Prof. Gebhardt)

3.3.3 Role of Middlemen

From the initial aim of the government resettlement program (*exploiting vast waste land and helping indigenes develop their economy)*, Kinh migrants have found out more promising chances to make a good living. They have flocked to the Central Highlands both to farm, and to work as traders in timber, cash crops, and they now dominate the main urban market (Human Rights Watch 2002, p.36). Many of them play the role of middlemen in the indigenes' economic system. The appearance of migrants coincided with the development of the market economy and commercial cash crops in the Central Highlands. This role of Kinh middlemen was not initially included in government plans but arose incidentally alongside the development of market-oriented economy.

More exactly, the weakness of indigenes created good conditions for the appearance and existence of middlemen. Middlemen generally fill up most of the shortages in the indigenes' farming working, including the process of producing, marketing, and transport. For the process of producing, the middlemen play the important role of offering cash loans or supporting investments (including fertilizers, seeds, irrigation), most of which the indigenous community have not used in their previous agricultural systems.

Moreover, as a result of the Kinh's historic background in the lowlands, they have accumulated skills as traders and are more familiar with the markets. The Kinh middlemen are therefore able to be an effective bridge to connect indigenous farmers and the market. The appearance of middle men is recognized as an essential link in the economic chain. They purchase agricultural products and transport them to the market. Even in remote villages, middlemen facilitate the producing activities of the indigenous people, as they understand the market demands well and can then inform the indigenous people about the changes in market demands.

In general, middlemen and indigenes both benefit from their relationship. However, the appearance of middlemen negatively impacts the livelihood pathways of the indigenous people, as indigenes inactively grow to depend on the supplements provided by the middlemen. Middlemen, in turn, earn from this middle position and have been known to take advantage of the weakness of indigenes. Although some farmers are dissatisfied with the way middlemen conduct their business and resent being treated unfairly, they still uphold this relationship. The strength of this relationship firstly resides on the capital dependence. Indigenes often owe middlemen for crop input. Subsequently, they do not enjoy the fruits of their crop output, but instead the middlemen do. After the indigenous farmers have paid their debts, they again have to lend cash for the next crop. This traps many households into a cycle of debt. Indigenes 'disadvantageous circumstances end up becoming worse, as the lack of capital is added by an absence of good roads and lack of infrastructure and vehicles. This all restricts information exchange and intragroup interactions (The World Bank 2009). Indigenous farmers are often not even aware of the market price of agricultural products, as they simply follow the arrangements provided by middlemen. If they react against the low prices offered by middlemen, they run the risk of not getting a loan for the next crops. "Suffering low price" (ép giá) is a typical term that indigenes use to describe their transaction with middlemen. Moreover, the reason why indigenes agree to sell their products for a low price is that traders often buy unprocessed goods (ibid: 41). Corn, cassava, and green coffee cherries are sold after harvesting, because indigenes *do not have room for shortage and have no facilities to process* (interview in Lac Duong District, in November 2014). Meanwhile, Kinh middlemen can arrange for primary processes and then store the goods until they can get a high price to resell. This is a reason why many indigenes blame middlemen for their poverty.

Mrs. Y. Hong, Kon Jo Ri Village, Kontum City

Indigenes are good-natured, credulous. They do not know how to negotiate. Moreover, their literacy level is low, they do not master calculation, and their Vietnamese language skill is not even good enough. Therefore, they just follow whatever the middlemen suggest. Debts and interest are usually not calculated transparently, but villagers are unable to argue. All the same, middlemen try to press price of agricultural products as low as possible. After harvest, indigenous farmers are left with a very modest amount of money after paying off their debts. They suffer hardship forever, forever.

Middlemen/traders are an essential chain in any economic system. However, the relationship between the Kinh middlemen and indigenous farmers is reason for concern, as it is generated based on an unbalanced power induced by unequal endowment in capital, infrastructure, and skills. The lack of power of indigenes places them in a dependent position and forces them to trade with the Kinh middlemen exclusively.

3.4 An Insight into Livelihood Pathways at the Household Level: Actor Role

The historical contexts, cultural patterns, and physical conditions at all three sites combine to paint a picture of the livelihood pathways of the indigenous people. Clearly, the way the indigenes have transformed their livelihood to cope with the constraints has varied from different groups at the same sites. The research results resonate with statements which argue that making decision is shaped by cultural, social, and historical factors and characteristics. Scoones and Wolmer state that "livelihood decisions are made within specific historical and agro-ecological conditions, and are constantly shaped by institution and social arrangements" (Scoones and Wolmer 2000, p.162).

However, some single cases at the study sites have unveiled that these statements above fall short of ascribing active roles to individual households. Although villagers share the same livelihood constraints and are together influenced by historical and social circumstances, some are more vulnerable or successful than others. "The different responses to insecurity, change are different between people," and "not all people live under the same conditions and create security in the same way" (Nooteboom 2003) are statements that underplay the role of context and increase individual determinants. Although social contexts and institutional processes have a strong influence on decision-making processes, these factors are not decisive enough to explain the dynamics of livelihood among households. Individuals, who are not only motivated by institutions and outside interventions but also have their own characteristics, play key role in forming their own livelihood pathways. When zooming in on household scale, we see that indigenes interpreted their specific sociocultural context in their own different individual activities.

The influence of the individual on the reproduction of social structure has been discussed through various concepts such as the actor (Wiesmann 1998; Bohle 2001) and theory of practice (Bourdieu 1977). In the specific realm of livelihood, scholars have examined this issue through the lens of style (Nooteboom 2003) or actor (Ulrich A, Ifejika Speranza C, Roden P, Kiteme B, Wiesmann U, & Nüsser M 2012; Roden P, Bergmann C, Ulrich A, & Nüsser M 2016).

3.4.1 Assets Approach

In order to shed light on individual cases in this research, the author considers behavior as a determinant of livelihood and applies the actor concept, which is interpreted by set of human assets to analyze decisions made by indigenes. Actor is seen as an active agent. And an action-oriented perspective requires a focus on the *meaning* assigned to an activity and the *means* used to perform the activity (Wiesmann et al. 2011, p.11). The term "meaning" refers to both material and nonmaterial assets, resources, and capitals which are assumed to reveal heterogeneity when we further investigate the personal circumstances of households at the study sites. In order to clarify this further, we look at a number of cases who successfully looked beyond their farm income or effectively intensified their crops. This approach bears some similarities to methodologies of "positive deviance" (Wells-Dang 2012, p.5). The idea is to look at a few individuals among a group who share the same resources but differ in a positive direction.

K. Plyng, an interviewee in Lac Duong District, is one of a very few villagers in his age, who had a chance to attend college. After graduating, he worked as a medical staff in his hometown and joined in cultural and arts activities. Later, he quit his jobs and set up the Gong team and then organized Gong performances as a tourism service. Now Gong performance, a traditional performing art, has become his livelihood. In this way, he can both earn a living and preserve his traditional culture. So far, there are 11 Gong groups in this area. Most of them were taught by Plyng. When we asked him why he chose Gong performances, he mentioned his love of culture and art as a main motivation. However, the way he described his choice reveals another clue.

> *I am the son of this land. My home town has a beautiful soul. Everyone recognizes its beauty, but they all just talk and no one does anything to develop it. I always worked at a state office, and had the chance to travel around. I saw that the tourism system in Lang Bian* [his hometown] *had not yet been highly developed, and had no certain orientation. I knew that if we wanted to develop tourism, we needed a certain tourism product to serve; otherwise, it is just like we beg tourists for food. This made me concerned so much. I would like to do something, but first I needed a proper vision and enough to make an adequate investment. I started by doing extensive research about our culture before organizing the first Gong performance* (interview at Lac Duong District, on 05.11.2014).

His proactive way of thinking distinguishes him from his fellow villagers. This sort of thinking led to the establishment of the first Gong performance group and then to a chain of Gong groups. While many people may join in with this activity due to their love of traditional culture, Plyng clearly laid the first foundations based on his knowledge about tourism, the market, his social relationships, and his strategic vision. His knowledge may not have directly influenced on his livelihood activities; however, it shows a good background and a wide vision to make his pathway change. Due to his knowledge and the achievement he gained when creating a new livelihood (not only for his family), Plyng earned the trust of villagers and was recognized as a village leader, even though he was not the oldest man in his village.

Even in traditional livelihood such as in agricultural activities, knowledge also plays an important role to encourage farmers to improve their work on farm and restructure their crops. The second significant case is Mr. Y T. Dran, leader of Le Village, in Lak District. His father was a wage earner for Bao Dai King (Wikipedia 2010); thus, he had the opportunity to attend school. He became a teacher in 1960 and worked as a research assistant at the linguistic department of the Ho Chi Minh University of Social Science and Humanity. Now he just works on his farm. Due to his background, he has earned respect of the villagers. His knowledge also supports him on farming. The progress of his work has resulted from the way of investing in crop inputs, deciding to restructure crops, and selling products. This man is one of the few farmers who does not need to loan money for crop inputs. Although his village traditionally specializes in rice cultivation, he is the pioneer in growing coffee. So far he has set up a stable coffee field. Moreover, he is smart enough to make some preliminary treatment of his coffee before selling, whereas most of farmers just sell raw coffee.

> *This area did not grow coffee. After reunification, we learnt it from the Kinh people. In my village, I was the first to grow coffee. Not only in developing coffee, but also in other crops, building house, etc., I was always the pioneer… I do not need to take out a loan for growing coffee. I explain to my co-villagers that if we loan money, we have to pay for both capital and interest after harvest. Then we cannot enjoy the fruits of our labor. I do not sell raw coffee. No matter whether the yield is high or low, I always keep it, dry and grind it. Then I store it and do not sell it until the price of coffee increases… Cultivating rice is easier than growing coffee. But we should not be afraid to risk investing in coffee, especially as rice crops do not always bring in profit, the price of rice is too low nowadays. So I would rather to develop coffee. Now I just keep one sào of rice, and rent the rest of my land. On my rice field I do not use any pesticide or fertilizer, instead I keep traditional techniques like we did in the past. I know these chemical substances are not good for people's health.*

Most of the good examples of cases about knowledge assets are middle-age or older people. The success of these old people results from their close-knit community and commitment to traditional values, regardless of livelihood (agriculture) or culture. Their decisions are based on knowledge, and their livelihood pathways are a result of accumulated processes. The knowledge involved here should be understood as both general knowledge and life experience.

In much research about rural livelihoods, young generations are expected to have a strong influence on households in terms of making decisions and undertaking new ventures, as they are more active; have more chances to access various information sources, in particular modern media; and are encouraged to contribute different ideas and enthusiasm for new strategies (Jordan et al. 2012). Unfortunately, young indigenous people at the study sites do not have great impacts, as they are facing a number of disadvantages. Due to the system of land being transferred to the following generations, the later generations have more land constraints to cope with. In particular, newly settled families are often in miserable situations, by virtue of shortage of material for their new lives. Moreover, the young indigenes nowadays are seen as a "halfway" generation. This is because they do not master their indigenous knowledge well and have not yet gained enough modern knowledge either. Many of them struggle to find a sustainable livelihood. During the field trip, we came across many young people who are not able to manage their new families well. This brings an impression that young people are not able to play a full role of an active actor to generate their livelihood pathways. The young generation, however, show strength through other aspects, since they own other assets.

Thanks to inheriting a new education system after the Vietnam War, young people have chances to work at state offices (e.g., as teachers, medical staff, or staff at local state offices). Even though it is a challenge for indigenous people to pursue study and strive for a stable job, this position brings some advantages that support them in their actor roles and in building up their household's livelihood pathway. First of all, a stable job means a stable income. When one family member works at a state office, this household has more capital for their daily lives and, more important, to invest to intensive crops.

Mr. Long is a rich and successful farmer in the Lac Duong District. He owns one ha of strawberry and is one of the very few indigenous farmers who is able to employ others. He hires both Kinh people and his co-villagers to harvest strawberries and work on his farm. In 2014, he spent *VND2.000.000 per day to pay for waged labor* (interview at Bang Gia Village, Lac Duong District, on 03 November 2014). This has great meaning not only in the economic effective of Long's work but also in his ownership power over his land. He states, *many villagers own a large amount of land, but they carelessly sell that land and then have to be waged earners. I will never sell my land, just buy more* (interview at Bang Gia Village, Lac Duong District, on 03 November 2014). He is fully aware of the development trend and the indigenous' deficiency. However, he seems to have been able to overcome those weaknesses and determine his household's livelihood pathway explicitly, since he has a stable capital condition, based on his wife's position.

> *Mr. Long, Bang Gia Village, Lac Duong District*
> *In this village, many Kinh people come and hire or buy land to grow flowers as flowers bring big profit. Our villagers now just grow vegetables and strawberries; no one develops flowers, because they lack capital and techniques. I have a plan to exploit the rest of my land for flower crops. My wife is a state employee; she can support me and invest the capital. Because of that, I set my mind at rest* (interview at Bang Gia Village, Lac Duong District, on 03.11.2014)

Positions like this that are strengthened by stable jobs (normally the state employment) could also bring indigenes important social capital in terms of quick access to development programs. Many stories from respondents indicate that those who hold a position at state office can access information from governments and other organizations such as NGOs more easily and quickly than the others. They, then, can have a more positive view and fully perceive about the implication and objective of such development programs. *My husband works at the Youth Union of this commune; last year he heard of a project in which the state was offering to invest for us to grow vegetables. I felt anxious about taking this risk, but he was fearless and bravely registered to join (interview at Da Nghit Village, Lac Duong District, on 03 December 2014*). People such as these are more likely to join government programs, in particular the programs that aim to improve people's economic status. Another case is Ma Khue, a Bana woman, living at Kon K'tu Village, in Kontum City. Her husband is a medical staff at his local commune. Due to his social position, he also plays a pillar role at home. Ma Khue shared us the story about their livelihood decision, which was more or less initiated by her husband:

> *When we got married, our life was bitterly tough. According to Bana's custom, a girl inherits its basic facilities from her parents to set up a new life. Because my family was too poor, I did not receive anything, but left everything for my younger siblings. My husband has been an orphan since he was young. We both started with "white hands". My husband strived for his study, and then worked at the local clinic. Once he was informed that the commune had a program, which offered loans to local people. First, my relatives dissuaded us from getting a loan; they warned us we would not be able to repay our debt. However, my husband was optimistic thought and encouraged me: Let's try just once. If we fail, we will never do it again. Then we used that budget to invest in a ginger crop.*

The profit from that crop helped us to afford to build our house (interview at Kon K'tu Village, on 22.12.2014).

Thanks to her husband's social position as a state staff, he can easily access information about government programs. With a full awareness of such preferential programs, he was more confident and optimistic about getting a loan from state and investing in his farm. His family is one of the rare families who make use of such opportunities; most other households in his village keep borrowing money from private shops.

Last but not least, a social position facilitates access to others. In comparison to the historically isolate circumstance of indigenes, working in a state office helps indigenes integrate with modern society and enrich their social relationships beyond their community, which positively contributes to their family's livelihood pathways. In particular, for some non-agricultural activities, relationships and interlinks play a special role and can even determine the success of investors. Relationships with outsiders also help indigenes learn and adapt to the commercial and market-oriented economy.

Ms. T. Selli was always a teacher. She is now a founder and leader of a Gong performing group. Her group was founded later than the others but right now attracts the most tourists. She admitted that her successful business was contributed by her relationships:

I used to be a teacher, but my salary was not enough to raise my children. I had a Kinh friend. Every time she visited me and observed groups in my village performing Gong for income, she encouraged me to found a group and start a business. At first, I would always find reasons to refuse it, and I thought there enough people who did this. Moreover, we did not have anything but an empty plot of land. We needed infrastructure, musical instruments, and many other investments to run Gong performances. My friend kept pressing on us and helped to invest in infrastructure. At the beginning, we did not have many guests. Again, my friend supported us. Since she had connections with some tour companies in Dalat, she introduced us to them. We gradually built on relationships, and have achieved prestige. So we have become more popular now (interview at Bon Dong Village, Lac Duong District, on 14.11.2014).

Actually, a social position as an office staff can be separately examined through different factors, such as capital access, social relationships, or information accessibility. However, considering the vulnerability context of indigenous people, it cannot be denied that apart from land and knowledge/experience as traditional assets, indigenes hardly own any other assets unless they hold a position at a state office, even if it is just a small position at a local commune.

Using ages to separate the two main assets that influence household decisions on livelihood may lead to misunderstandings. Actually, age does not present a holistic view in this case. The issue discussed above should be summed up as the preliminary assets determining livelihood pathway being knowledge-experience and the social position gained by jobs. Technically, the former asset is linked more to older people, while the latter refers more to the young generations. These assets cannot be exactly assigned to two those age groups. However, after a long time of practicing self-subsistence and being isolating from the outside, the younger generations nowadays have more chances to access assets beyond their community to support them in making a living. For their livelihood, young people tend to internalize the values of modern society more.

3.4.2 Non-Assets Approach

Besides using certain assets to interpret the decisions of indigenes, we realize that there are cases which seem to be shaped by nonlinear actions "without either explicit reason or signifying intent" (Bourdieu 1977, p.79); the factors that help indigenes arrive at their decisions are not always clear. We call this approach a non-assets approach. It actually borrows similar ideas from the habitus[2] concept. In these cases, the process agent interprets history structure into his actions which may be an unconscious part, since these preconditions are already inveterate in him. Therefore, individual agent practices sometimes accept an explanation based on recent motivation, as an agent's characteristic (part of "yesterday's man") is underlay. This study recognizes the role of the subjective actor in schemes of perception, conception, and action but does not address the factors or assets forming these actions; instead, the products of actors will be recognized as a part of process.

Mr. Them is one of the Gong performance organizers at the Bon Dong Village, in Lac Duong District. He organizes Gong performances as a livelihood in addition to his farm work. In comparison to the other Gong groups, his group is outstanding, thanks to generous investments made in the group's infrastructure (performing lot, musical instruments, sound system, etc.). Originally, he only worked as a farmer, and his profile was not different from that of the other villagers. His success is clearly explained by his adventurous characteristics and high motivation to make a good living.

> *Mr. Them, Bon Dong Village, Lac Duong District*
> *To improve life, it is necessary to know how to invest. In general, our community does not master this issue. In the past, I was so venturesome when I got a loan from the bank. Thanks to contacts with the majority Kinh, I learnt from them and thought if they could do it, so could I. I took the chance to grasp new things from this modern society and was willing to cope with challenges. In 2005, I invested to develop flowers, and then I changed to do business in tourism. I did not have any money at all. I accepted a loan from the bank, which motivated me to work hard to earn a profit from my business. I have now already paid off my debt. My co-villagers judged my decision as a reckless. They have never held a large amount of cash in their hands, how could they live with such a big debt, they would be filled with panic of debt. Several million VND for growing coffee is even too big for them, let alone several hundreds of million VND for flower producing or tourism developing...*

His different livelihood pathway at present is a "product of a chronologically ordered series of structuring determinations, the habitus" (Bourdieu 1977, p.86). Obviously, Mr. Them is not only affected by preconditions but also dominated by his earliest experience and his disposition. During his livelihood transformation process, he has coped with failure several times. He *began with flower crops, and*

[2] Habitus is a historical product and embodied subjective motivation. "The principle of these individual differences lies in the fact that, being the product of a chronologically ordered series of structuring determinations, the habitus" (Bourdieu 1977, p.86). In practice, the habitus is history turned into nature. The "unconscious" is never anything other than the forgetting of history, which history itself produces by incorporating the objective structures it produces in the second natures of habitus. In each of us, there is part of yesterday's man, yet we do not sense this man of the past (ibid: 79).

then invested modern facilities in a karaoke shop. Unfortunately, a bad storm that year swept all away and destroyed his house and his karaoke shop. He did not give up but only grew more motivated and determined to begin again. This time he started a Gong performance group. He got a loan to invest in Gong team and has worked on developing it since (interview at Bon Dong Village, on 09 November 2104). Mr. Tham has accumulated experience in running a business, and his livelihood has been improved further. It should be noticed that within the process of shaping his livelihood, "rules and values of social space has become internalized via routine interaction and observation" (van Dijk 2011).

Another case is Mr. Niu, a Jarai man who lives at the Plei Bur Village, in Kontum City. He is famous for his success in developing agriculture in a highly diversified and intensified manner. He stopped growing annual crops (cassava) and then diversified into perennial crops including coffee, rubber, and pepper. He explains his livelihood pathway as follows: *I do not grow cassava anymore, because the land is no longer fertile and output is too low. I just copied from the Kinh people. I observed the Kinh growing their crops and then followed them. I grow coffee, rubber, and pepper. No one in my village has grown pepper before* (interview at Plei Bur Village, Kontum City, on 29 December 2014).

It is unclear why Mr. Niu has succeeded in developing intensive agriculture, whereas other farmers in his village have not. Using habitus to describe the operation of agents' action allows us to clarify the history structure, which is sometimes obscured from our eyes but still exists and indirectly constitutes actions. As a soldier during the Vietnam War, Mr. Niu lived outside his village for a long time and traveled to different places. During this time, he learned many new things from outside his village and beyond the traditional norms, which he had grown up with. This may not explain his success nowadays; however, it cannot be denied that his history partly makes him more open and quick-witted to learn and reconstruct his livelihood pathway. In another aspect, Bourdieu explains that habitus, which results from a homogeneity of conditions, is what enables practices to be objectively harmonized without any conscious reference to a norm (Bourdieu 1977, p.80). This is also true in this case, since there is no antecedent in his community of people developing intensive perennial crops like pepper, and he does not have any experience in this area himself either. His distinctive pathway is "shaped by the habitus, but this pathway in itself also contributes to the shaping of the habitus" (Jordan et al. 2012, p.289) and has become the pattern for his next activities.

We have applied both the "structure" and "actor" concept to shed light on the factors that simultaneously shape livelihood pathways. An individual agent's role can normally be assessed by certain indicators as assets. However, as the dynamic of actions reflects on the dynamic of the livelihood pathways, it is difficult to reach a fixed set of factors embedded in individual livelihood activities. In addition, it cannot be denied that livelihood pathways are not determined by mono-causal factors but rather are constituted by the processes of an individual actors adjusting to a certain context. These arguments along with the results above, therefore, prove that we need to "bridge the gap between the structural and individual dimensions" (Nooteboom 2003) when examining how indigenous communities make their living.

3.5 Conclusion

3.5.1 Dynamic Livelihood Pathway

Heterogeneity in each household's livelihood strategy when responding to the same external conditions reflects the dynamics of livelihood pathways in the Central Highlands and reveals the role of the individual agent (Fig. 3.5).

In general, agriculture still occupies a major position in their livelihood typology, and on-farm income accounts for the largest share of household income. Even, implications of diversification are nothing more than to enhance and support agriculture. Therefore, the indigenous people's livelihood pathways are assessed as not being particularly dynamic. Indigenes still weigh survival strategy against a way out of poverty (Moreda 2012).

Figure 3.6 below gives a simple illustration of the stages through which indigenous people transform their livelihood. This figure allows us to sum up and generalize the choices related to making a living in three study sites over time. Even though indigenous households might not absolutely practice through all stages of livelihood transformations described in the figure, the advanced order of stages is apparently ruled based on the availability of essential assets, which enable households to move toward the more value activities. The level of livelihood dynamics presents different pictures at three study sites.

This scale shows that Lac Duong generally reaches the most advanced livelihood stage since indigenous households in this area are relatively more active in joining in with more high-value economic activities. Migration has not been included in the figure because it is an aspect that can appear at any stage in the livelihood typology. The last noticeable point of this figure is the absence of industrial activities. Indigenes do not engage in the industry at all, not even in the food-processing industry. According to the three-sector theory, the economic activities of indigenous people have skipped the secondary sector, as they make their living mainly in the primary (agriculture) and tertiary sectors. Remarkably, from among the very few nonfarm activities, tourism shows up as a popular activity at all three study sites. Tourism is the most advanced stage of rural economic activities. Once participating in tourism

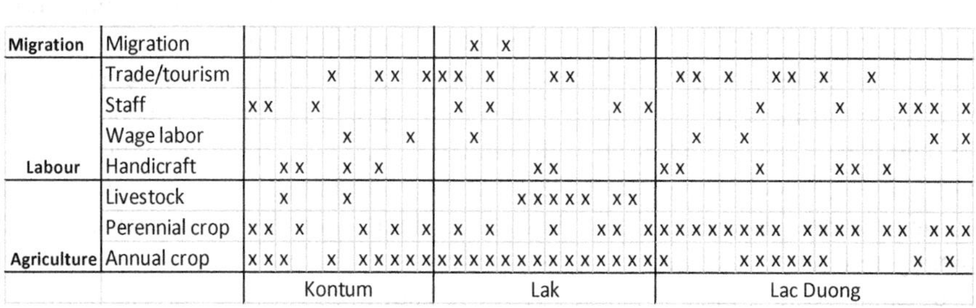

Fig. 3.5 Livelihood dynamics at study sites (*Source*: Interview at Kontum City, Lak District, Lac Duong District from October to December 2014)

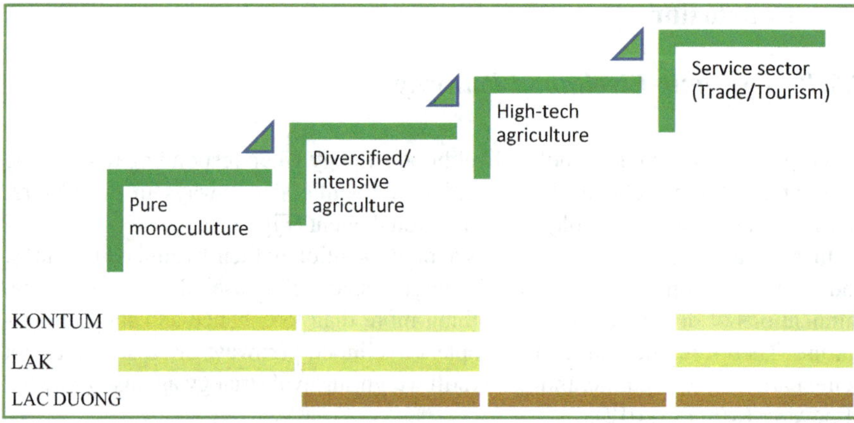

Fig. 3.6 Transformations of rural economic activities in the Central Highlands (Note: The hue (from lightest to darkest) displays different levels people get involved (from lowest to highest) in each stage of livelihood transformation) (*Source*: Huỳnh Anh Chi Thái)

development, indigenous people begin working in service-based enterprises. As they struggle with land constraint context, tourism is expected to bring in an extra income for indigenes and will be analyzed as a resilient livelihood in the next chapter.

3.5.2 Factors Forming Livelihood Pathways

The process through which indigenes design their livelihood (or their livelihood pathway) is discussed in this section as the sensitivity aspect of indigenous communities and is examined through a compound lens comprised of interrelated dimensions of contexts, institutions, and assets.

The research results unveil that although indigenes heavily depend on the land or, in other words, that land is recognized as a pillar in their livelihoods, land constraints do not directly affect local livelihood pathways and are not the only factor that influences a dynamic livelihood pathway. Technically, the influence of land and other factors on indigenous livelihoods can be sketched out in Fig. 3.7, though it is not easy to determine why household chooses a particular strategy. Under land constraints and the encouragement of off-farm income, poor households are expected to be more enthusiastic in engaging with diversifying their livelihoods to compensate for their deficit. Yet, in reality, well-off families appear more often in nonfarm activities. Although poor households have economic motivations, this motivation is not strong enough to activate them, as they also struggle to cope with the challenges of owning few assets (capital, skill, knowledge, etc.). Likewise, young people are assumed to have the most dynamic livelihood pathways. However, in reality, this is not always true. As land is divided through inheritance, farm size is becoming increasingly smaller for younger generations. Young farmers usually own little

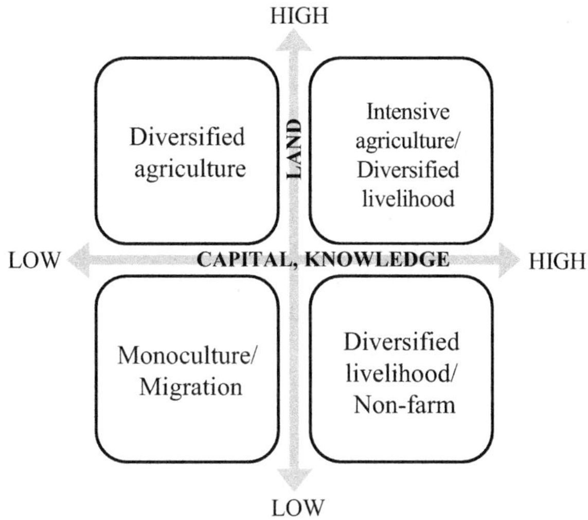

Fig. 3.7 Correlation between land and other factors determining livelihood pathway (*Source*: Huỳnh Anh Chi Thái)

land – the most significant asset. In addition, if the shortage of land is not discouraging enough, the lack of capital puts, even more, pressure on their burdens, as this means, they cannot afford to apply high technologies and develop intensive crops. Meanwhile, older households seem better able to cope, since they have accumulated assets over a longer period.

Nonetheless, livelihood in this study is not presented as intentionally evaluated outcomes or final results, rather the process through pathway approach. The more well-off households might start their business with the same assets as poor families. Moreover, a holistic people-centered focus has been used in this research, and we have downplayed some quantitative indicators, such as age, income, poverty line, etc., and then highlighted the existence of household assets and the ways indigenes mobilize them into their livelihood pathways. Using this approach, we found a significant relationship between development pathways and assets of social capital and human capital. An individual agent designs his/her livelihood pathway more or less based on social structures, which are illustrated through historical, economic, and cultural characteristics. In addition, this study proves, as does some other research on rural livelihood, that "household assets determine household productivity" (The World Bank 2008).

Among assets, capital and education/knowledge are the most important that affect indigenes' potential to engage in high-value livelihood regardless of whether that is of a non-agricultural or agricultural nature. Among four archetypes, Greener (2002) distinguishes the following: "reflexive-agent as subject agency," "non-reflexive agents as subject agency," "reflexive-agent as object agency," and "non-reflexive agent as object agency." The "reflexive-agent as subject agency" groups present full role of actors, since they are well enough aware of social structures and are able to engage in strategic behavior to improve their livelihoods (van Dijk 2011). If indigenes have capital and/or an open mind with a stable knowledge background, they are more likely to be categorized into the "reflexive-agent as subject agency" groups. Knowledge which might be gathered through experiences facilitates the production of effective crops. Meanwhile, education teaches indigenes to apply new technology to agriculture or to diversify their livelihood and access high-value jobs. Most of the successful cases in the developing livelihood pathway are the households that have at least one state staff member. However, in general, the gap in education attainment between indigenes and other ethnic groups (especially the Kinh majority) remains large. Along with land constraints, low access to education and a lack of capital are great contributors to a less dynamic in the livelihood pathways of indigenous people.

References

Bohle H-G (2001) 'Vulnerability and Criticality'. IHDP Update 2/2001. Newsletter of the international human dimensions programme on global environmental change, pp 1–6

Bourdieu P (1977) Outline of a theory of practice. In: Gellner E et al (eds) Cambridge University Press, Cambridge

Brooks N (2003) Vulnerability, risk and adaptation: a conceptual framework, Tyndall Centre Working Paper, p 38

Chambers R, Conway G (1991) Sustainable rural livelihoods: practical concepts for the 21st century. Available at: http://opendocs.ids.ac.uk/opendocs/handle/123456789/775. Accessed 22 June 2015

Dang HA (2010) Vietnam: a widening poverty gap for ethnic minorities. In: Indigenous peoples, poverty and development. World Bank, Washington, DC

Dang TH, Shively G (2008) Coffee boom, coffee bust and smallholder response in Vietnam's Central Highlands. Review of Development Economics 12(2):312–326. https://doi.org/10.1111/j.1467-9361.2007.00391.x

Dang NA, Tacoli C, Hoang XT (2003) Migration in Vietnam. A review of information on current trends and patterns, and their policy implications. In Regional conference on migration, development and pro-poor policy choices in Asia. Dhaka, Bangladesh

Füssel H-M (2007) Vulnerability: a generally applicable conceptual framework for climate change research. Global Environ Change 17:155–167. https://doi.org/10.1016/j.gloenvcha.2006.05.002

Greener I (2002) Agency, social theory and social policy. Critic Soc Policy 22(4):688–705

Hickey GC (1964) The major ethnic groups of the South Vietnamese Highlands. The Rand Corporation, Santa Monica

Hickey GC (1967) The highland people of South Vietnam: social and economic development. The Rand Corporation, Santa Monica

Human Rights Watch (2002) Repression of Montagnards: conflicts over land and religion in Vietnam's Central Highlands. Human Rights Watch, New York/Washington, DC/London/Brussels

Hussein K, Nelson J (1999) Sustainable livelihoods and livelihoods diversification, 69. Brighton Institute of Development Studies, Brighton

IFAD (2009) Understanding power and processes. Guidance notes for institutional analysis in rural development programmes

Jordan GNM et al (2012) Resilience, Pathways and Circumstances: Unpicking livelihood threats and responses in the rural Philippines. Durham University, Durham

Kaag M et al (2013) Poverty is bad: ways forward in livelihood research. Radboud Repository. doi:https://doi.org/10.1063/1.3033202

McElwee P (2008a) Becoming socialist or becoming Kinh. In: Duncan CR (ed) Civilizing the margins: Southeast Asian government policies for the development of minorities. NUS Press, Singapore

McElwee P (2008b) "Blood relatives" or uneasy neighbors? Kinh migrant and ethnic minority interactions in the Trường Sơn Mountains. J Vietnamese Stud 3(3):81–116. https://doi.org/10.1525/vs.2008.3.3.81

Moreda T (2012) Linking vulnerability, land and livelihoods: literature review. International Institute of Social Studies, The Hague

Nooteboom G (2003) A matter of style: social security and livelihood in upland East Java. Radboud University Nijmegen, Nijmegen. https://doi.org/10.1063/1.3033202

Prowse M (2003) Towards a clearer understanding of 'vulnerability' in relation to chronic poverty, 24. Chronic Poverty Research Centre, Manchester

Roden P, Bergmann C, Ulrich A, Nüsser M (2016) Tracing divergent livelihood pathways in the Drylands: a perspective on two spatially proximate locations in Laikipia County, Kenya. J Arid Environ 124:239–248

Scoones I (1998) Sustainable rural livelihoods: a framework for analysis. Available at: http://mobile.opendocs.ids.ac.uk/opendocs/handle/123456789/3390. Accessed 15 July 2015

Scoones I, Wolmer W (2000) Pathways of change: crops, livestock and livelihoods in Africa. Lessons from Ethiopia, Mali and Zimbabwe. Institute of Development Studies, University of Sussex, Brighton

Swift J (1989) Why are rural people vulnerable to famine. IDS Bull 20(2):8–15

The World Bank (2008) Agriculture for development. Washington, DC: The International Bank for ReconstructionandDevelopment/TheWorldBank.doi:https://doi.org/10.1596/978-0-8213-7233-3.

The World Bank (2009) Country social analysis: ethnicity and development in Vietnam

The World Bank (2012) Poverty reduction in the Central Highlands. Analysis of Ethnic Minority Labor-related issues

Turner BL et al (2003) A framework for vulnerability analysis in sustainability science. Proc Natl Acad Sci U S A 100(14). https://doi.org/10.1073/pnas.1231335100

Ulrich A, Ifejika Speranza C, Roden P, Kiteme B, Wiesmann U, Nüsser M (2012) Small-scale farming in semi-arid areas: livelihood dynamics between 1997 and 2010 in Laikipia, Kenya. J Rural Stud 28:241–251

UNHCR (2006) Vietnam: situation of indigenous minority groups in the Central Highlands. UNHCR

United Nations (2010) Internal migration: opportunities and challenges for socio-economic development in Vietnam. Ha Noi United Nations Viet Nam, Ha Noi. https://doi.org/10.1038/1811574d0

United States Agency for International Development (2008) Vietnam Central Highlands needs assessment. USAID

Valerie K (ed) (2014) Well begun but not yet done: progress and emerging challenges for poverty reduction in Vietnam. International Bank for Reconstruction and Development/The World Bank, Washington, DC

van Dijk T (2011) Livelihood, capital and livelihood trajectories: a more sociological conceptualization. Prog Develop Stud 11(2):101–117

van Dillen S (2004) Different choices: assessing vulnerability in a South Indian Village. Verlag für Entwichlungspolitik, Saarbrücken

Warner K (ed) (2006) Perspective on social vulnerability. UNU-EHS, Hohenkammer

Watts M, Bohle H (1993) The space of vulnerability: the causal structure of hunger and famine. Prog Human Geogr 17(1):43–67

Wells-Dang A (2012) Ethnic minority development in Vietnam: what leads to success? Background paper prepared for the 2012 Poverty Assessment, Hanoi, April. Hanoi, Vietnam

Wiesmann U (1998) Sustainable regional development in rural Africa: conceptual framework and case studies from Kenya. University of Berne Switzerland, Institute of Geography, Berne

Wiesmann U et al (2011) A Human actor model as a conceptual orientation in interdisciplinary research for sustainable development. In: Wiesmann U, Hurni H, with an international group of co-editors (eds) Research for sustainable development: foundations, experiences, and perspectives of the Swiss National Centre of Competence in Research (NCCR) North-South. University of Bern, Bern, pp 231–256

Wikipedia (2010) Bảo Đại. Available at: https://en.wikipedia.org/wiki/Bảo_Đại. Accessed 1 Mar 2016

Wisner B et al (2003) At risk second edition: natural hazards, people's vulnerability and disasters. Routledge, London

Chapter 4
Livelihood Resilience – A Case Study: Community-Based Tourism (CBT)

Abstract This chapter examines community-based tourism (CBT) – one of the pathways indigenous people choose to cope with livelihood stress – through the lens of resilience livelihood. Social capital is a significant asset for vulnerable groups to develop CBT and a good example of resilience, in that CBT shows how the transformed social capital maintains its role in facilitating collective actions. Bonding connections and bridging networks are applied to assess the implications of social capital for community participation in tourism activities. The results describe three stages of CBT development at three study sites in which social capital contributes to the formation of different tourism management styles: private business, cooperative business, and community enterprise.

Keywords Community-based tourism · Social capital · Community participation · Livelihood resilience · Tourism management style

4.1 Introduction: Transition from Farm to Nonfarm and Community-Based Tourism

Since the livelihood strategy of the indigenous people has broadened to more than purely crops and livestock production, nonfarm activities have become an indispensable trend and a positive pattern of response to land constraints. One of the most promising nonfarm activities is tourism, in particular community-based tourism (CBT), as mentioned in the previous chapter. CBT is expected to be a great potential when it becomes more developed in the Central Highlands and is expected to enhance the liquidity of the indigenous people. So far, the reaction of the indigenous people to overcome livelihood constraints has been analyzed from a sensitivity aspect, which means their livelihoods are evaluated as being less dynamic and lagging behind in comparison with other groups. This section will discuss the last component of vulnerability concept – resilience – to examine to what extent indigenes as an active agent could mitigate the negative effects and resist the exposure.

According to the characteristics of CBT, resilience will first be assessed through the resources that a community owns, the assets they can access, also referred to as

H.A.C. Thái, *Livelihood Pathways of Indigenous People in Vietnam's Central Highlands*, Advances in Asian Human-Environmental Research,
https://doi.org/10.1007/978-3-319-71171-3_4

their buffer capacity. A livelihood is thus resilient if it can absorb the impacts of disturbances and maintain its key functions (income, poverty reduction, etc.). In the specific context of the Central Highlands in Vietnam, which is exposed to constraints of material assets (land and capital), the study considers immaterial assets as a means to compensate and support communities in overcoming external pressures. Two of these immaterial assets, human and social assets, play a significant role in the development pathway. Moreover, these assets are particularly significant for discussing CBT, which is defined as "a tool to strengthen the ability of community organizations that manage tourism resources with the participation of the local people" (Suansri 2003, p11). The livelihood resilience perspective in this study is, therefore, examined in close connection to social resilience. In addition, as a service activity, tourism also calls for a synthetic analysis of other assets and capitals. Moreover, the second indicator that should be tested is self-organization. Self-organization encompasses various levels of autonomy, the freedom to act, collective action, self-help, self-reliance, independence, power, and control of one's own action (Ifejika Speranza et al. 2014). Thus, the analysis of self-organization closely relates to and again is based on the buffer capacity of human and social capital. In this way, it addresses how a community participates and makes a living on their own.

4.2 Tourism Resources and the Reliance on Own Resources

The study was conducted in two districts and one provincial city where tourism has gradually been developing. These three study sites – Lac Duong District, Lak District, and Kontum City – share common characteristics that contribute to shaping a joint tourism product branded as "Central Highlands' tourism," which is characterized by exotic minority cultures and stunning mountainous landscape. However, each site also has its own specific characteristics and experiences different paths of forming and developing tourism. Lac Duong District, which is located near Dalat City, Lam Dong Province, is a resort center in the highlands created during the French colonial period and is the most popular tourist destination so far. Next, Lak District, which is a part of Dak Lak Province, known as the center of this mountainous area, embodies distinct cultural identities of the Central Highlands. The last one is Kontum City, a former administrative center of the French colonialists, which is the last of the three to start its development as tourism destination. When assessing CBT through the lens of livelihood resilience, we look at how the indigenous people mobilize assets to develop tourism. Tourism resources are one of the most important assets in this regard, a prerequisite for developing tourism products (Table 4.1).

Table 4.1 Tourism status in Vietnam's Central Highlands

	2000		2005		2010		2010
District/city	International arrivals	Domestic arrivals	International arrivals	Domestic arrivals	International arrivals	Domestic arrivals	Share %
Kon Tum	1292	15,265	4055	27,786	58,500	68,900	3.31
Dak Lak	6520	137,765	14,540	188,609	50,000	450,000	13.00
Lam Dong	71,000	654,000	100,600	1,460,300	200,000	2,600,000	72.78
Central Highlands	**84,376**	**850,313**	**129,130**	**1,863,902**	**335,000**	**3,512,400**	**100**

Source: Institute for Tourism Development Research (2010)

4.2.1 *Tourism Resources*

Regarding tourism resources, it would be superficial if we just listed attractions promoted in brochures by tourism companies. For this reason, we discuss what local people can access to practice tourism (entitlements) in addition to the resources they may own (endowments). Moreover, since CBT is a product based on community, we need to identify authenticity as an indicator to separate "front" and "back." The model of "front" and "back" was introduced by MacCannell (1999) to address the issue of authenticity. The front refers to what is obviously presented to the tourist; meanwhile the back implies the authenticity of the local's private lives (Dolezal 2011). We use the "front" concept for the available tourism services and "back" for what lies behind the scenes, the untouched area that reflects real life and is not included in tourism products. However, it is difficult to separate front from back, and MacCannell warns that "these are sometimes transformed one into the other" (MacCannell 1999). In fact, some "shows" are based on the structure of reality, and some "back" objects are not intentionally concealed.

Tourism resources are one of the main factors shaping the different images and status for each study site. Every study site has its own tourism potential. In order to develop CBT, it is also important to assess the authenticity derived from their daily lives and the degree to which they can access and exploit resources.

Lak District

Tourism resources are conventionally divided into natural resources and cultural resources. Lak District is famous for its beautiful landscape, which includes the Lak Lake and the great biodiversity. Lak Lake and the cultural landscape are encompassed by a mountainous area. Lak Lake is the second biggest lake in Vietnam and one of the most scenic spots in Vietnam. Located among a mountainous area, the lake brings fresh air for this site. Therefore, it has been exploited as a tourism potential for a long time (Table 4.2).

Another location in the Lak District, the Chu Yang Sin National Park, has not yet officially become a tourist attraction. This natural park is under the control of the

Table 4.2 Tourism resources in Lak District

		Authenticity		
Endowment	Entitlement	Front	Back	Description
Lak Lake	x	x		Scenic spot, fresh air
				Site for elephant riding and dugout rowing
Chu Yang Sin National Park		x		Primeval forest ecosystem, Buon beep waterfall
				Under management of Forest protection department
Cultural space of Gong	x	x	x	An oral and intangible masterpiece recognized by UNESCO.
				Historically performed in traditional festival
				Is modified to include in tourism program
Elephant riding	x	x		A value property used as vehicle to transport timber from forest in the past
				Used to drive tourist across Lak Lake
Dugout canoes	x	x		Originally, a vehicle connecting villages surrounding Lak Lake
				Vehicle for tourists who visit villages on the other side of Lak Lake
Housing architecture	x	x	x	Originally a stilt house made of natural materials
				Is now modified, using industrial materials
				Offered as accommodation for tourists
Agricultural landscape	x		x	Cultivating paddy
				Catch fish with net

authorities. The regulations for conducting tourism in forests are even stricter, in particular when it involves foreign tourists. Due to some ambiguous political reasons, foreign tourists are hindered from reaching national parks. A limited number of foreign tourists manage to visit the forests, and these are all accompanied by staffs of tourism companies. However, it is not free and legal for the indigenous people to access forests to conduct CBT.

Apart from these state-managed areas, forests in general have been degraded. Local people are clearly aware of the tourists' attraction to natural landscapes and are disappointed with the deforestation in their hometowns. They blame the immigrants for the forest scarcity.

Mr. Ma Va, M'Lieng Village

In the past, this district was full of forests, always covered by trees. Just 10 years ago, the forests were lost. Now we do not have forests anymore. People from the North (Cao Bang, Lang Son Province) came here and destroyed everything to plant cassava and maize. The authorities did not know about this... Jun Village used to be a center for welcoming tourists. Tourists were sent to our village since we had such magnificent forests. Western tourists love nature so much. Now those Western tourists no longer need tour guides, they can go by themselves, and save their money. They are not afraid,

> *because there are no longer thick forests around. There is no place, where there are no people – the Nung, Thai, Tay, Muong*[1] *– they are everywhere, even in deep forests.*

Besides natural resources serving as the potential for ecotourism, culture also plays a significant role meeting the increasing demand among tourists to experience the real life of host communities. Particularly for CBT, community culture serves as a key asset. Nowadays, ranges of tourist products that more or less involve local people are sold as CBT in Vietnam. However, the true value of a CBT product is anchored in the authenticity which is revealed from the real life, rather than a simple "show" (MacCannell 1999). Therefore, we will next discuss the existence of cultural value and "front" and "back" model.

The indigenous people in the Lak District are M'Nong – one of the main indigenous groups in the Central Highlands. So far, some of their cultural identities have been integrated into tourism services, such as the Gong performance, elephant riding, and dugout canoes. All activities are based on the traditional customs of the indigenous people; however, some have been heavily modified and are no longer truly authentic. Originally, Gong was performed in traditional festivals. Now the indigenous people at this study site do not hold as many festivals as before, and so the Gong is rarely practiced. Thanks to tourism, they still perform Gong regularly but have to compose new melodies to suit the tourism activities. In addition, elephants were used in the past as a vehicle to transport timbers. Now the indigenous people use them to carry tourists. Apart from these available services, many traditional customs have, in fact, gradually disappear (Figs. 4.1 and 4.2).

> *Mr. Ma Va, M'Lieng Village*
>
> *Many traditional customs are gradually being lost. For example, Gong – we only put on performances when the authorities request it. In the past, we played Gong regularly. Now we cannot keep that up. We are much busier. Who will work on farm if we are playing Gong? Before we had more free time, there was only one crop per year. When I was a child, we did nothing in this month after the harvest, just feasted all day and played Gong. In this modern society, even though we work year in year out, we cannot catch up the other groups...*

However, there are also aspects of the villagers' daily life that appeal to tourists and that the local people are not even aware of. For example, daily activities not organized into tourism products can attract tourists who wish to observe the authenticity of exotic and minority cultures. The daily activities are interpreted by many tourists as a pristine beauty without any pretending. The M'Nong villages are located not so far from the main road and so are easy for tourists to access, but they are also far enough away to separate them from the modern town to allow the villagers to lead a tranquil life. M'Nong people have been gathering at the Lak Lake for decades and here uphold some of their traditional customs. The local people still live in stilt houses with thatched grass roofs and bamboo walls. Although material from forests to build traditional houses has become rare and the M'Nong people have to use industrial material as a substitution, they prefer to keep the traditional

[1] Nùng, Thái, Tày, and Mường are ethnic minorities who originally lived in Northern Vietnam and who migrated to the Central Highlands after the Vietnam War.

Fig. 4.1 Russian tourists and tourism activities at Lak District (*Source*: Huỳnh Anh Chi Thái)

Fig. 4.2 Sunrise at Lak Lake, Lak District (*Source*: Prof. Gebhardt)

architecture of stilt house. More notability is the conventional ways they uphold to make their living. Common images here that attract tourists include those of local farmers cultivating their rice fields with the rudimentary equipment, local fishermen dropping nets in the early mornings, and oxen and buffaloes being herded along the village paths. These sketch a picturesque overview of the authentic life here. "Real life, authentic details are woven into the fabric of our modern solidarity alongside the other attractions" (MacCannell 1999), and in this case the peaceful real life of the indigenous people contributes as an asset for them to develop CBT.

Lac Duong District

While the Lak District is well known for the peaceful landscape of Lak Lake, paddy fields, and the more or less preserved traditional community space, Lac Duong is a modern rural area. The local people have experienced an evident transition in their producing activities as well as in their lives. Most potential tourism resources in this district have already been exploited and transferred into tourism products. Natural and cultural resources have contributed to form Lac Duong tourism brands for some decades now, and during this process, tourism products have gradually become more commercialized, and their authenticity has increasingly faded. Here again the disparity in the chances of accessing tourism resources between indigenous people and outsiders is worthy of attention.

Lac Duong District is recognized as having many advantages in tourism development. Located adjacent to Dalat City, Lac Duong District shares the same weather conditions with Dalat City, which is well known for its fresh mountain air all year round. Thanks to the benefits of the microclimate, this destination draws tourists who enjoy cooler climates and wish to escape the heat and humidity of the tropical weather in South Vietnam. In particular, during summer time, the climate is the first comparative advantage in tourism development.

In addition, Lac Duong District has many tourism resources. One of the most important attractions in Lac Duong is Lang Bian Mountain (Vietnam Online 2015), which is located on Lam Vien Mountain and consists of the two highest peaks: Lang Bian (2.167 m) and Bidoup (2.287 m). Lang Bian Mountain is famous for its spectacular nature, as well as its marvelous views over the whole wide area of Dalat City, the forest, the rivers, and the village of indigenous people at the foothills. The mountain has been designated as a national park and is managed by the Forest Protection Department. Differently to the case of the national park in Lak District, this national park has had permission to develop tourism for a long time now. So far, there are two tourism business units in this area: the Lam Dong Joint Stock Tourism Company and the Center for Ecotourism and Environmental Education of Bidoup-Nui Ba National Park (Fig. 4.3).

Apart from the spectacular natural landscape of this mountain, Lac Duong District has another stunning landscape, the lake of Gold Stream. The Golden Valley Resort was established at this site in 2005 and has become a well-known attraction. It is managed by the Lam Dong Water Supply Joint Stock Company. According to the master zoning plan (No. 97/2003/QĐ-TTg, 22.07.2013) (Lam Dong Province 2009), this site is designated as one of the four national tourism zones, which are supposed to be given priority in terms of investment and development. However, up to now no successful project has been launched in this area. Similar to Lang Bian Mountain and two other resorts in Lac Duong, the tourism resources are mainly exploited by the Vietnamese. A few indigenous people are hired in to assist.

The indigenous people in Lac Duong District are K'Ho. Their notable cultural identities which have been modified have now become an explicit tourism product. One of these is the Gong performance. Differently from Lak District, the indigenous people in Lac Duong have completely stopped playing Gong in their daily lives, except in Church services for holy masses. These days, communities are required to apply for certification from the Department of Culture, Sport and Tourism if they want to organize Gong performances as a tourism service. It is ironic that the K'Ho need a verification from outsiders (Vietnamese official offices) for what is part of their cultural identity. Because Gong performances are seen as a service, the indigenes can possibly cope with asset shortages by translating this culture value into tourism activities. Besides certifications issued by the state, which only 11 groups have been granted to now, financial capital is also required to organize Gong performances. In fact, K'Ho do not have a common space for Gong; households need to invest in constructing facilities to perform; and moreover, they not only play traditional musical instruments but also combine these with modern ones, which also require large investments.

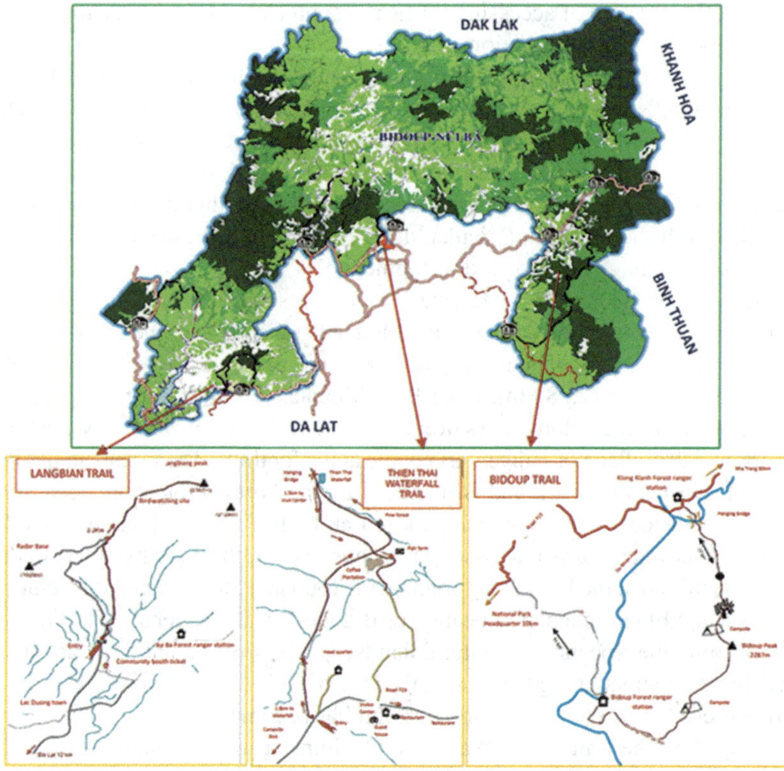

Fig. 4.3 Overview map of Bidoup-Nui Ba National Park (*Source*: Bidoup-Nui Ba National Park Center for Ecotourism and Environmental Education 2014)

> *Mrs. Sela, Bon Dong Village, Lac Duong District*
> *The Gong performance now is so different from the traditional one. Especially the dance, which we now compose following a modern style. If we stick to the original dance, our performances would not draw as much interest. This also goes for the musical instruments. Besides traditional musical instruments, we also use electronic musical instruments. The songs also, some are sung in our language, others are mixed with Vietnamese. No one is interested in our traditional style performances except researchers and journalists, who want to learn about our culture.*

Another cultural resource which should be mentioned here is traditional handicrafts. The K'Ho group in Lac Duong comprises two subgroups: Lach and Chil. The Lach are well-known for their Gong performances, but the Chil, who have not practiced this custom for a long time, have stuck to weaving as their traditional handicrafts. The Chil's traditional handicrafts is now commercialized and mainly used by outsiders. In order to create better conditions for earning an extra income, the Chil are allowed to sell their weaving products at the Lang Bian resort without paying taxes. However, their business is facing challenges of competition with industrial products. The young generations are not optimistic about the chances to promote their traditional handicrafts as a livelihood and lose their motivation to

Table 4.3 Tourism resources in Lac Duong District

Endowment	Entitlement	Authenticity Front	Authenticity Back	Description
Lang Bian Mountain	–	x		Developed and managed by lam dong tourism joint stock company and Bidoup-Nui Ba National Park
				The indigenous people have chance to join in tourism from the Center of Ecotourism and Environmental Education, Bidoup-Nui Ba National Park
Gold stream	–	x		Managed by lam dong water supply joint stock company
				Local people are hired for running errands
Gong performance	–	x		Need certifications from the Department of Culture, sport, and tourism to organize gong performance as a tourism service
				Have to invest in constructed yard and modern musical instruments
Traditional weaving	x	x		Used mainly by outsiders
				Cope with competition with industrial products
Agricultural landscape	x	x	x	Modern, joyful scenery of a productive and diversified agricultural activities

Note: *X*, available; –, not fully able to access

learn weaving. The Chil also import industrial weaving products and mix these with their traditional handicraft for selling (Table 4.3).

In addition to these explicit tourism resources, the specific landscape of the indigenous people's daily life should also be considered as another asset in Lac Duong. In comparison with the other sites, Lac Duong has gain the best agricultural outcome and is described as a productive and vital area. Thanks to their economic achievements, they have gradually been able to adjust their daily life to become more modern, meaning they increasingly lose their customs and traditional lifestyle. It is difficult now to recognize their unique culture and distinguish it from ethnic Vietnamese. The only distinguishing thing left is their language; the appearance and lifestyles can no longer help to identify them.

Ms. Bluy, Bon Don Village, Lac Duong District

Our village has changed more and more, and is becoming more modern now. When the landscape changes, we need at least the tilt houses to reflect traditional space. But now the living demands of villagers are completely different from the past. This is in comparison with Lac Village,[2] where the local people have preserved their ethnic character well. JICA[3] suggested that we build a tourism village like in the north. Nevertheless, it is difficult…

[2] Lac Village is a rural area of the Hoa Binh Province in the northwest region of Vietnam. This village is over 700 years old, and the scenery of Lac village attracts many tourists. In 1993, it received permission to become a tourist destination. At present, there are 20 homestays at this village (Le 2015).

[3] JICA stands for Japan International Cooperation Agency. Some years ago, this organization funded for a 4-year project in Lac Duong District for "Strengthening Community-Based

However, scenes of productive agricultural activities create a bright and joyful picture to attract visitors. In addition, coffee, strawberries, and various vegetables are grown here, making these fields a potential agricultural tourism site that may appeal to visitors who are interested in learning about diversified agricultural activities in the uplands.

Kontum City

Although the last study site we visited – Kontum – is a provincial city,[4] the villages we spoke with here are the poorest as well as the most traditional in comparison with the other villages in Lak and Lac Duong District.

The most notable tourism resources in Kontum City are the authenticity. CBT in Kontum is on the preliminary stage, with the most famous communities joining the tourism industry being the Kon K'tu and Kon Jori. These communities are separated from the city center by Dak Bla River. They were isolated until the Kon K'lor Bridge was built to connect these villages with the urban city areas. However, up to today this area is still rather isolated from the other modern areas. This has created conditions for the indigenous people to preserve their traditional customs. The Kon K'tu Village, as its name means "far and desolate village," is located at the end of the road and has become the most popular CBT destination in Kontum City.

The villagers still live in stilt houses surrounding the communal house named Rong house, and they still cultivate traditional crops such as cassava, maize, rice, and sugar cane. These all contribute to shaping a peaceful rural landscape. Some cultural identities have already been translated into authentic tourism services, such as Gong performances and dugout canoes. The tourism products are similar to those in Lak District, except that they are far more authentic, in particular in the way they perform the Gong. In Kon K'tu Village, the Gong is always performed in the evening, outside, and in front of the Rong house, and there are normally about 20–30 villagers who gather here to perform. They perform numerous passionate folk songs and traditional dances, which are taught to next generations. Moreover, although a Gong performance costs a fix price, the Gong group enjoys gathering together with many villagers, even though the benefit is smaller because Gong group has to share them with more people. In this way, they preserve their specific cultural features.

Moreover, this area is characterized by a romantic natural landscape, as it is located alongside the poetic Dak Bla riverside and encompassed by the forest. In the past, this area was an ideal place for the Bana people, who preferred to lead a peace-

---

Management Capacity of Bidoup-Nui Ba National Park" (http://www.jica.go.jp/vietnam/). In one of the JICA programs, the indigenous people had a chance to visit some CBT destinations in the north including Lac Village (mentioned above) to learn from the experiences there.

[4]In Vietnam, the provinces are divided into districts (*huyện*), provincial cities (*thành phố trực thuộc tỉnh*), and district-level towns (*thị xã*) which are subdivided into commune-level towns (*thị trấn*), or commune (*xã*). A commune-level town can be upgraded to a district-level town, and district-level towns can develop into provincial cities. In 2009, Kontum district-level town was upgraded to provincial cite.

Table 4.4 Tourism resources in Kontum City

Endowment	Entitlement	Authenticity		Description
		Front	Back	
Dak Bla River	x	x		An attraction explicitly described in tourism product, contributing to the beautiful landscape
				Offers dugout canoe service along the river
Poor forest	x	x		Trekking is a popular tourism service
				Forest has been destroyed and becomes less attractive
House architecture	x	x	x	Retain traditional house architecture including residential house and collective house
				Lifestyle within traditional house is not always revealed
Gong performance	x	x	x	Following the traditional gong performance
				Truth festival space of gong performance gradually lost
Handicraft	x	x	x	Basket weaving, brocade weaving
				Less developed and not usually included in tourism program

ful life, away from fortune and modern life and competition and fighting with other tribes.[5] This partly attributes to their characteristics, as they are seen as the friendliest group in this area. The cultural landscape is also shaped by handicrafts such as brocade and basket weaving, albeit to a lesser degree (Table 4.4).

4.2.2 *Other Resources*

Beside different specific tourism resources, tourism in the Central Highlands is also influenced by other conditions. These are generally similar between study sites and have partly built up the tourism characteristics for this area.

Position

Tourism is a specific activity, which brings consumers to the productions. Ease of access is one asset that helps in developing tourism. For CBT, location plays an even more important role. The location of the community should not only meet the requirement of easy access but also be considered to be at a "just enough" distance.

[5] According to a Bana legend, in the early ages, the tribes in this area always fought each other over properties and power. But the Bana sons disapproved of the war and did not want to join in the fighting anymore. They then decided to settle down in peace nearby the Dak Bla River. This legend also explains the name of their village: Kontum – a village of lake. Since then, they have been known for their peace-loving nature (Kon Tum Province 2009).

On the one hand, this distance should be far enough to keep a community separate from the city center and its influences and far enough to draw the interest of tourists. On another hand, since CBT is developed on a small scale, the distance from one community to the other destinations should be near enough to allow for easy enough access and links to main tour area.

In general, all three study sites are located at a near enough distance, as they have easy connections to the main tourism hubs. The Lac Duong District is located 12 km north of the center Dalat[6] and 333 km north of Saigon[7] on the National Highway 20. Meanwhile Lak District is in the very south of Dak Lak Province, 67 km from Buon Ma Thuot[8] City and 226 km to Nha Trang[9] City on the National Highway 26. And Kontum City is located in the north of the Central Highlands, relatively far from the south tourism hub but still with easy connections to Danang[10] on National Highway 14 (292 km to the North) (http://maps.google.com). Among them, Lac Duong District is the most popular destination so far, which partly results from its good position of being relatively close to the biggest tourism hub in the south and next to the tourism hub of the central of Vietnam (see Fig. A.3 in Annex).

Regarding the "far enough" distance, the communities in Kontum City have the best advantage. They are 12 km away from the city center and are separated by Dak Bla River. Meanwhile, Jun, Le (in Lak District) and Bang Gia, Bon Don, and B Nor'C (in Lac Duong District) are located right in the middle of the commune-level town and are thus able to get more contacts and learn more from Vietnamese ethnic. The changes in their traditional customs are more or less reflected by their villages' positions.

Political Condition

The study sites all share the same common political condition as the rest of Vietnam. As in most of Vietnam, this area is a safe and friendly place to visit. And unlike some big cities, violent crimes and even petty crimes are rare.

Recently, there has been some unrest in this area due to the minority dissatisfaction with land issues and religious freedom (United States Agency for International Development 2008). The authorities assume that the protests are the work of "outside forces" who are bent on disturbing internal unity and national security (UNHCR 2006). They have thus become more cautious about foreigners. Some villages in the

[6] Dalat is capital of Lam Dong Province and has long been a resort center in highlands (Wikipedia 2005a).

[7] Saigon is the biggest city and the big tourism center and business, financial hub in Vietnam (Vietnam Tours 2016).

[8] Buon Ma Thuot is the capital city of Dak Lak Province and is recognized as capital of Central Highlands as well.

[9] Nha Trang is a coastal city and capital of Khanh Hoa Province, one of the tourism hubs in Vietnam (Wikipedia 2005b).

[10] Da Nang is a municipality and identified as a tourism hub on south-central coast of Vietnam (Wikipedia 2009; Vietnamnet 2015).

Central Highlands are no longer permitted for foreigners. Lak District, Lac Duong District, and Kontum City are, however, areas where the authorities allow for more open political condition and where they promote the development of tourism.

However, given the complicated political condition in the Central Highlands, it cannot be denied that there are certain limitations to developing tourism in this area and that these require careful considering.

Mr. Plyng, Bang Gia Village, Lac Duong District
The security agency always thinks that all foreigners are spies, are enemies. Therefore, they do not allow us to run a homestay, in particular in the Central Highlands. In the Northern Highlands you can... They call the Central Highlands a sensitive area.

4.2.3 Reliance on Own Resources

The tourism resources discussed above allow local people to reply on their own resources to develop tourism. Obviously, the contemporary tourism programs that companies offer at these sites are mainly based on the indigenous people including their culture and their supplies.

Beside accommodation, traditional food, and other available services, the cultural identity of the indigenous people's daily life is one of the most valuable resources that contributes to building up a true-to-life CBT product. This also allows them to maintain their identity, without the need to commercialize their culture into a tourism show. Although these communities share common cultural characteristics, each site has its own individual strength and offers different tourism products. Lac Duong District has diversified tourism products that combined ecotourism and CBT, whereas Lak District offers various community tourism services. And Kontum City is known as a sleeping princess, with its authentic cultural landscape.

The market for this kind of tourism seems to have strong potential. Many outbound tour operators in Europe and America are increasingly showing business interest and are opening more tour CBT programs to attract travelers in search of culturally richer experiences, with closer contact with locals (Dolezal 2011; García Lucchetti and Font 2013; MacCannell 1999). These sites with their exotic and minority cultures promise to meet the demand for authentic experiences and become attractive destinations.

Although the indigenous people can rely on their tourism resources, the degree of promoting and exploiting these tourism resources is still modest. There are other resources that have been neglected, such as agricultural tourism in Lac Duong District and ecotourism in Lak District. Also, the level of community participations to transfer these resources into tourism products is influenced by different factors. These will be discussed in the following section.

4.3 Human Capital and Willingness to Participate

Human capital in this study is considered in terms of household size and number of laborers, level of education and language skills, and knowledge of tourism that potentially contribute to developing CBT.

4.3.1 *Human Capital*

Household Size – Number of Laborers

In the context of land constraint, the large household size pushes the local people further toward exposure to the vulnerability context, since they have to divide their land up into more plots for the many family members. Big families, however, have more labor capacity to diversify their livelihood strategy. When the labor in a household can undertake the tasks of farm working, they can assign other tasks to other family members. In this way, farmers are sometimes able to manage to join in with tourism activities, even during the busy crop seasons.

Mrs. H'Be, Le Village, 12.2014. Labor allocation into agriculture and tourism

We have to divide the work between the family members. For instance, there are five persons in our family. Some of them will work on the farm; others do tourism-related activities. We cannot abandon our farm. It is our life... My son has just completed his bachelor's degree, but has not yet found a job, so he stays at home and helps me with the farm work for now.

Up to now, agriculture has always been the major source of income of indigenes. And as discussed above, indigenous households are rather reserved in transforming their livelihoods. They recognize that tourism can offer an extra source of income, but only households with "extra-labor" capacity can join in with tourism activities.

Level of Education and Language Skills

Although a service sector like tourism tends to require a higher education level than agricultural production, education level does not have a direct influence on fostering or hindering the development of CBT. It does, however, indirectly determine the opportunities indigenous people have to join in with tourism activities. A higher education level gives farmers an open mind and teaches them to be quicker to adjust from the agricultural producing sector to a service sector like tourism. Level of education reflects the quality of labor and is also closely linked to knowledge on the language and tourism.

Since the indigenous people have their own language, not all of them can communicate in Vietnamese; knowledge of English is then even more of a challenge for them to welcome foreign tourists. Due to the history of Vietnam as a French colony, there are some indigenous elders who can speak French well. They evidently become delegates and are employed in important positions in tourism activities. In some cases, human capital can compensate for the lack of the other assets. So far, most local people who join in with tourism activities own at least some tourism resources (either material or immaterial resources), for example, a beautiful traditional house, elephants, or dugout canoes, or have the ability to perform Gong. Others, who are well-educated and master a foreign language, for example, have the chance to become tourism operators, meaning community tourism activities do need to depend on a tour guide or a translator from companies either inside or outside their village (Van Breugel 2013).

Mrs. Mi Ly, Le Village, Lak District

I learnt at a French school when I was young, so I can speak French well. I have joined in with tourism activities because I like practicing my French... Meeting tourists and speaking this language makes me happy. I have been working at this office since 1997. Now I am getting old, I am about to retire, but there is no one to take my place... Some local people can become tour guides, but they cannot speak a foreign language, and they do not have the chance to learn. They are just able to say some simple words such as hello, how are you? Where are you from? And they cannot introduce the attractions to the tourists.

Knowledge of Tourism

The tourism industry calls for specific diverse skills and labor to meet the demands of the tourism industry and cope with the fluctuating circumstances of the business world. However, official training is certainly not the only source of knowledge. According to our observations, most indigenous people never attended any tourism courses. There are two main sources through which local people can learn about tourism. Firstly, some groups who have already joined in with tourism activities as suppliers of tourism services, such as Gong performances, Dugout canoes, or elephant riding, get the chance to have contact with tourists and learn the basics of how tourism works. However, they are hired to passively join a certain part of the tourism industry, and they are not highly motivated to engage more in tourism.

Mr. Ma Lu, Jun Village, Lak District

Until now, there is no one in our community who has thought to start a tourism company themselves, never. Those who own an elephant or dugout canoes will join in with tourism activities. If someone in our community can develop their own business, of course would be glad.

The second group consists of people who are fortunately trained before they enter into the tourism business. For example, in Le Village, there is a project to develop CBT, supported by Global Competitiveness Facility (GCF) and operated by Van Phat LTD Company. Their aim is to organize a set of training courses to provide local people with basic tourism knowledge and skills. Although the trained people are then assigned to different service sectors, they have gotten a more comprehensive picture of tourism and have become more confident about expanding their tourism services in the future. In other words, knowledge of tourism is closely related to the willingness of a community to participate.

Mrs. H'Be, Le Village, 12.2014. Tourism knowledge as a human asset

I have been trained in many skills such as first aids, cleaning house, welcoming and communicating with tourists. It is not so difficult. When tourists stay at my home, they are very pleased because of the tidiness, and the tourists compliment on my cooking. This makes me very happy. When I have enough money, I will invest in building up a new tilted house to offer even more accommodation, and I will run a tourism business on my own.

4.3.2 Willingness to Participate in Tourism

The level of community participation in the tourism is particularly low in developing country (Tosun 2000). There are a set of external factors such as operational and structural limits that inhibit the community participation, and people tend to exhibit greater persistence in the developing world than in the developed world (ibid). From a local perspective, the willingness of villagers is seen as an internal factor, which supports the determinant of community participation. The willingness to participate in tourism depends on the level of their knowledge of tourism and on which impacts (advantage or disadvantage) they know about this industry.

Most of the interviewees promptly mentioned the economic tourism benefits and the lack of land as determinants fostering them to join tourism activities. The livelihood constraints push them to join tourism activities, since tourism as an extra source of income acts as a pull factor. Many villagers do not think that tourism is as hard as working on a farm. However, this willingness is limited to passive participation to be hired and be told what to do. Local people in an undeveloped tourism area express the desire to join tourism activities but wait for an active trigger from outsiders rather than taking initiative themselves.

Mr. Ma Va, M'Lieng Village

We would be glad to have tourists visit our village, like in Jun Village. We do not have many skills, but if the tourists stay in our homes, we can earn money... Between investing in tourism and planting coffee, I think tourism is better, because tourism is not as hard as working on the farm. But until now, no one has come here to open a tourism company for us.

Another motivation for the participation in tourism of indigenes is the respect for their own traditional culture. Indigenous people believe that tourism activities can result in recognition and preservation of their customs and cultural value. The preservation of culture is not weighed as highly as economic driving force, but at least the village elders do recognize it as a benefit and express their pleasure when tourism can contribute to reviving their traditional customs.

Mrs. Mi Ly, Jun Village

At the beginning, we did not totally trust in tourism activities. Particularly the young people, who had never had contact with the foreigners before, could not imagine what western people would look like. Later, they gradually came to know more about tourism. We understand that tourists are not dangerous, they enjoy visiting us, and we can earn money from dugout canoes and elephant service they use. Especially, we perform Gong for them. This also means the youth can learn to play Gong again. Otherwise, our young people nowadays would not know how to play, they are not as interested in Gong anymore, and they prefer to play with their smartphones, listen modern music...

However, there are also some factors that hinder the villagers from willingly joining tourism activities and running tourism businesses themselves. The first relates to the lack of basic knowledge about the tourism industry. As mentioned above, knowledge plays a role in a community's willingness to participate in tourism. However, knowledge also has contrary effects. In some cases, the more people know about how complicated it is to run tourism business, the more hesitantly they become. Most interviewees in M'Lieng Village, who have never before participated in tourism, expressed a strong willingness to do so. In contrast, the villagers in Jun and Le Village were reluctant to run tourism activities themselves, because they have witnessed how the Vietnamese run tourism activities in their village and can clearly image the difficulties and challenges.

More specifically, local people are not fully aware of how meaningful CBT is and that it is expected to "be managed and owned by the community, for the community, with the purpose of enabling visitors to increase their awareness and learn about the community and local ways of life" (Suansri 2003). Therefore, they all weigh up their challenges on the lack of finances over the assets of their community themselves – one of the most significant elements to build up CBT (including their culture, their local ways of life, etc.). In fact, the Kinh people have been borrowing from the indigenous culture to run tourism programs, and tourists visit villages to learn about their culture as well.

Ma Lu, Jun Village, Lak District

If we want to develop tourism, the first and the most important thing is finances. It is so difficult. Where can we borrow money for tourism? Where are our resources? No one can do it. Moreover, you need to be good at communication. Finance and communication are prerequisites.

The last constraint is the fluctuation of the tourism market which makes local people hesitate to join in with tourism activities. Tourism has been developed in this area for decades. Local people can follow the fluctuation of the tourism market year after year. No matter what the reason for the instability in the tourism market, it is something out of their control, and the fact that the number of arrivals has decreased recently makes them unwilling to engage in tourism (Table 4.5).

Table 4.5 Tourism arrivals in Lak District in the period 2008–2011

	2008	2009	2010	2011
Number of arrivals	13,155	9557	8875	8820
Domestic	5560	3652	3,740d	4250
Foreigner	7595	5905	5135	4570

Source: Lak People's Committee (2014)

4.4 Role of Social Capital and the General Model of CBT Development

"As well as having its own intrinsic value" (DfID 1999, p14), this study regards social capital as one of the five core assets of sustainable livelihood framework and as one of the three main assets in developing CBT as a resilient livelihood. Social capital has a direct impact on other types of capital and may be particularly important as a resource of last resort for the poor and vulnerable group (ibid, p.14). The role of social capital in the tourism context, as well as in other fields, all relates to the necessity of mobilizing collective actions. Social capital plays an even more significant role for CBT, which is an economic activity based on and for the community.

Most tourism research tends to "only touch the very surface of the concept of social network" (Hwang 2012, p.50) by assessing connections within networks, measured using proxies such as the number and types of groups of which an actor is a member and the degree of participation in groups and networks (The Department for International Development 1999; Ifejika Speranza et al. 2014). However, since our study sites are rural areas and tourism has not yet been highly developed here, measuring of social capital based on simply counting the number of (formal) groups is not likely to yield much. In the present study, social capital is considered through two components: structural and cognitive (Harpham et al. 2002; Jones 2005; Claiborne 2010). While the structural component derives from associational links, activities, networks, rules, and roles, the cognitive aspect refers to norms, value, reciprocity, and trust (Jones 2005). "At the simplest level, these two components can be respectively characterized as what people "do" and what people "feel"" (Harpham et al. 2002, p.106). However, the structural and cognitive components will not be analyzed separately but will be examined simultaneously to gain more insight into social capital in terms of quality value such as trust, reciprocity, and connectedness within communities.

Another important thing is that the connectedness type of social networks is categorized into bonding and bridging (Chen et al. 2012; Claiborne 2010; Hwang 2012; Jones 2005). Bonding entails within a community and bridging involves extra community relations (Jones 2005, p.307). To view indigenous people from a perspective beyond their historical isolated position, their social capital will be examined in a wider and open network of both connectedness types.

4.4.1 Social Transformation and the Bonding Social Capital

Networks, Common Rules, and Norms

Similar to other rural areas where reciprocal networks and social networks are based on kin and place of origin, and on more recently formed local networks, the indigenous people in the Central Highlands of Vietnam are also characterized by their community networks. However, this characteristic has been changed and unfolded differently between study sites. The persistence of traditional norms and its mixture with new rules varied due primarily to socioeconomic conditions.

Historically, the villages (also called bon, plei, and kon in their languages) were the smallest administrative units and played an extremely significant role in spiritual and daily life of indigenous people (Hickey 1967). Even a person's individual identity was defined by his/her village, as they had no strong awareness of individual ego (Nguyen Ngoc 2008). Village leaders played an important role in maintaining the law and connecting villagers in strong solidarity (Hickey 1967). Their society was controlled by an unwritten customary law which was passed down through the generations by the village elder headers and was respectably followed by most villagers. Therefore, communities' self-organization was strong based on social coherence,, and traditional norms instead of individual interest.

The massive interventions through policy reforms of the Vietnamese Government after 1975 caused the disorder in indigenous societies and generated "voices arguing that more respect and sensitivity should be paid to the importance of cultural traditions among minorities" (McElwee 2008, p.201). That had first made traditional norms and rules, in some sense obsolete (United States Agency for International Development 2008); subsequently, structural social capital described through the hierarchy of social relations has collapsed. Politically, the villages became a state unit, controlled by new rules. The role of the elder became less important as his responsibilities were mainly taken over by state-appointed administrators. In addition, as a result of the state plan of resettlement and the spontaneous migration prompted by the appeal of profitable crops, the intermingling of Kinh people with indigenous people has fragmented their village space, which resulted in a less coherent village structure leading to a fragmentation of its functions. The ties in indigenous communities are now clearly based on more than their internal kinship. However, even kinship, which is described as "an amalgam of matrilineal descent and inheritance, clan-based residence" (Salemink 2003, p.33), has weakened nowadays. Since the customary law has been made obsolete, clans have lost their roles in their kin groups and no longer feel fully responsible for preserving their traditions or regulating their society. In fact, their societies are increasingly being transformed from village-centered to family-centered. The connections within the family are replacing the village community as the center of their social network.

Relation of Trust

Another facet of social capital is the trust which is seen as laying at the heart of a community (Thammajinda 2013). Trust is a positive basis for collective actions, which is attributed by the norm and beliefs mentioned above and is also influenced differently by the transformations.

Among our study sites, Lac Duong experiences the most transformation that pushes this town toward progress. These study sites have been designated as a township, and its villages have been reformed as the modern villages with a new title of "population group" (*tổ dân phố*), where the community structure is not identified anymore. The strong state interventions have pushed back their customary law that has subsequently weakened the connections in the village. It is now difficult to identify the community ties in the modern living space, as the Kinh are intermixing more and more within the indigenous communities. Villagers often mention trust as a value in assessing their community networks and as a key element in exposing the shift happening in their society.

Mr. Plyng, Bang Gia Village, Lac Duong District
In the past, we never had dispute. Villagers relied on nature, just planted paddy fields and maize. It was just enough to live on, every people had enough for their lives. We bred livestock freely on the fields, and left belongs outside, but were never afraid of anyone stealing them. Since the Kinh have come, everyone has had to build fences, and we cannot trust people anymore. The past was so nice, when we had more freedom and people behaved more civilized.

Trust is an important indicator of social capital, and since the local people have lost their trust to each other, "it is most likely impossible to enable to gain a collective agreement in community" (Thammajinda 2013, p.53). Collective activities in these communities are rare and completely absent in their tourism activities.

In a contrast scenario, it was unanimously agreed in Kontum City that communities adhered to the rules very well to preserve their trust. Study sites in Kontum appear rather isolated and well-bounded maintaining a traditional village structure. These villages are of thatched single-family houses arranged around a central communal house. The common house is an important symbol of community-respective groups for centuries as the spirit of the village. Elders' voices still weight in the communities' daily life and solidarity. The respect and adherence to these traditional norms were central to their continued functions thus far. Interview information supported the idea that norms and trust facilitated collective actions. Villagers cited examples of mutual labor, which they have been arranging in farming (harvesting, peeling cassava, drying, and cutting cassava) and in daily life. The special locations as well as the certain moderate development stage that has not brought much changes to the society are the positive conditions preserving strong ties of communities in Kontum study sites.

At a more developed stage, the contribution of social capital in Lak District has been more convincing. Inevitably, villagers at Lak District admitted to the loss of cohesion in their communities in comparison to the past. "In the past we freely shared labour and willingly helped when a family had a wedding or funeral, but now only relatives help each other, and it is not entirely free" (interview at Le Village, Lac District, on 01.12.2014). Local people, however, maintained their traditional lifestyle (living in stilt houses) and livelihood (primarily cultivating paddy) and still had a stable basis for their collective actions toward tourism development.

The erosion of social capital among villagers was revealed at all study sites and could be attributed to various reasons. First, since land had been seen as the most significant possession in the traditional society, changes in land use and land management inevitably had deep impacts on indigenous lives. Lands had always been common properties never officially allocated to villagers but, instead, managed by elders. After the state implemented land policy reforms, individual land ownership was instituted. As a consequence, villagers developed the perception of individual ownership. This partially broke the village collective spirit and brought about the pursuit of individual, rather than collective, benefit. Second, urbanization and development made sense in terms of opportunities for easy access and adaption to the customs and habits of the Kinh majority. Although Kontum[11] is a (provincial) city, the study villages at this site were less developed than the ones in some other districts.[12] Furthermore, due to the specific isolated locations, the study sites in Kontum tended to keep themselves peaceful and separated, rather than exposing their culture and close-knit network to be diluted by outsiders, like in Lac Duong and Lak District. Finally yet importantly, the appearance of migrants (nothing more than the consequence and/or driving force of urbanization and development) is stated as the third factor in the erosion of social capital. Whereas there were a few migrants in the study villages of Kontum and Lak, the number of Kinh people who had settling among the villagers in Lac Duong made this village structure unrecognizable and greatly weakened their social ties.

Given the vulnerability context with which indigenous communities were coping, bridging social capital brought a great deal of support in the form of resources and networks from outside. Of those external connections, the relationship between communities and tourism companies was the most important. These relationships developed differently between study sites, depending on the management type of the companies, and will be discussed in the next section.

4.4.2 Bridging Social Capital and the Three Development Stages of CBT

In addition to maintaining community networks, bridging social capital is also an important aspect of encouraging communities to engage in tourism development (Thammajinda 2013). In particular, for rural communities, who often lack capital and business skills and are just starting to develop tourism, outside networks provide a great deal of support. If the connection discussed above refers to horizontal involvement (bonding social capital), the participation and connection of indigenes with organizations outside the neighborhood constitute a vertical engagement (bridging social capital). Of those vertical engagements, the relationship between communities and companies is the most important bridging network that should be highlighted.

[11] In Vietnam, the province is divided into districts *(huyện),* provincial cities, provincial cities (*thành phố trực thuộc tỉnh*), and district-level towns (*thị xã)* which are subdivided into commune-level towns (*thị trấn*), or commune (*xã*). A commune-level town can be upgraded to a district-level town, and district-level towns can develop into provincial cities depending on their urbanization level.

[12] This can be understood, since processes of urbanization in Vietnam are sometimes not corresponding to stage of development (World Bank 2011). And the disparity of development pace in every administration unit might be caught.

Although tourism was managed in many different ways at every study sites, each site had one dominant type that represented the feature of community participation as a reflection of its social capital in correlation with different levels of developing tourism. Villages at three study sites are experiencing three different CBT development paths. An insight into these paths will shed light on the determinant role of communities' social capital (role of innovators and ability to control an equal ownership distribution) in relationship with outsiders/middlemen. The contributions of different actors should be weight to find out a sustainable cooperation in which the role and the right of each actors are clearly delineated.

Private Business Run by Outsiders

Study sites in Kontum appear to be the first step of development model where tourism is just at the beginning stage. These communities stood out because of their traditional networks, which supported villagers in getting agreements for working in tourism. Villagers were attracted to tourism activities by the opportunity for self-reliance and by the idea of showing off their culture to others while benefitting their communities.

Ms. Ma Khuong, Kon K'tu Village, Kontum City

People sometimes ask me why I work in tourism. Well, because the tourists love my village so I feel I should. It is not just for me, but for my community as well. If it only brought benefits for me, I would definitely not make the effort. But the money we earn is kept in a community fund, so I have more motivation to join in tourism. We use those funds, for example, to renovate the common house, and to celebrate festivals in the new common house. We are Catholic, so the community fund is also used to renovate the church. If someone is ill, we can also send her/him some of that money…

However, as the last site to develop tourism, such activities here were relatively limited and mainly run by outsiders. At the time we carried out our fieldwork, Highland Eco Tours (Highlands Eco Tours 2010) was the only company running at Kon K'tu Village. Impressed by the long-standing community networks and understanding its role in a CBT program, the company owner had no way better than building his business on a partnership with the community that adheres to and respects community customs at this first development stage. In turn, villagers are eager to participate in his tourism activities. They leased out their land to build homestays and in this way got a monthly income, even during the off-season. They had also set up a group to perform Gong music, offered dugout canoe tours, etc. (Fig. 4.4). Villagers were able to share in the profits of tourism, and a small part was reserved for community funds.

Mr. Huynh, Vietnamese Tourism Company's owner, Kontum City

I run a CBT business, so the community is my partner. I cannot promise that CBT will help alleviate their poverty, but it certainly contributes a small part to their community fund, and makes them happy. The first step when I established a CBT business in this community was to meet with the village patriarch, who thoroughly grasps the village's status and whose voice is important. Then I held some village discussions to ask what they would and would not like. The most important thing is the network with the villagers, because the people here find sentiment more important than money. They are not highly

Fig. 4.4 Gong performance at Kon K'tu village, Kontum City (*Source*: Huỳnh Anh Chi Thái)

disciplined, but they participate in tourism activities because they enjoy it and like me, not due to money alone.

Tourism activities at this area present an image of one product, one community. At the moment, there is, on one side, no conflict between households when they work together, and they prefer to develop a unique CBT product for all. On another side, villagers expressed their loyalty in cooperating with the company. It sounds ideally, as the definition of CBT seems to come into practice in this case. However, some may argue that tourism in Kontum is still in its initial stage that the tourism revenue is not high enough to trigger the competition and conflicts. Furthermore, there is not any community organization taking charge of developing CBT. Community networks and kinship alone do not help local people self-organize tourism. The success of tourism in their village still depends much on management from outsiders. No egoism and no organization and high loyalty to tourism entrepreneurs are the features of this case that shows a sign of a very early step of CBT development.

Cooperative Business

At a more developed stage of Lak study sites, the community networks, which have been degraded in some sense, are performing their role though. Social capital of community is still stable for their collective actions toward tourism development and strong enough to raise the local voice when cooperating with outsiders developing CBT.

The Duc-Mai cooperative is one of the first tourism businesses established in Lak. The initial management board of this cooperative, which worked to mobilize local assets for tourism and offered jobs for villagers, was founded by one Kinh and three indigenous people. Thus, the villagers have many opportunities to be involved. In this cooperation, local people play a distinction role, which cannot be substituted. The villagers are providing different tourism services depending on their available assets, which are specialized into homestays, elephant riding, dugout canoes, and Gong performances. A complete tourism product, then, is a result of mobilizing all the various contributions from different households and presents a good example of cooperative business. This diminishes the conflict between groups, and even more, their social capital is seen as a "dependent variable" (Jones 2005), as it is consolidated through tourism activities.

Moreover, study cases of Lak District are lively example of social transformation in the Central Highlands. Since the traditional community networks have been eroded, kinship comes to play a new role in facilitating collective activities. Community networks, therefore, shifted from village-centered to family-centered. When participating in tourism activities, local people prefer to share job opportunities with their relatives first. In Lak, Gong set is a prestigious item belonging to a clan, so the local people joining in gong performance groups were mainly relatives. In addition, the mutual trust among kin is strong enough to purchase a common property, which was then used as a community asset in tourism activities. Villagers in Lak District also cited the example of their clan collective efforts to pool their money and buy an elephant, so the young men in that clan could share the responsibility of providing elephant rides for tourists. Kinship was expressing a new picture of community networks.

On another aspect, the bridging social capital building with tourism companies brought a great deal of support. The company in this case plays a central role to mobilize separate services from single households into tourism products and managed and assigned duties to each supplier based on tourists' demands.

Tourism activities were the catalyst that enriched and consolidated connections between households. However, due to the lack of relevant expertise and knowledge, local people expressed their anxiety about "*taking responsibility for the documents needed to manage a tourism business, such as asking for business permits, reporting tourist arrival numbers, tracking tourism revenue, and submitting income tax forms to the authorities*" (interview at Le Village, Lak District, in 12.2014). Because of this anxiety, they let the Kinh member take the lead in handling all such procedures and designing and operating the tour programs. In fact, this business unit was no longer exactly a cooperative, but more like a family company, as it was currently being managed by family members. Thus, the implementation of a CBT program likely meant negotiating mutual interests between communities and the Kinh owner. When discussing the sharing of benefits, local people admitted that they ever claimed for the higher profit sharing. Yet the long positive interaction and trust between villagers and the company had the effect of cushioning disagreement. Villagers, after that, considered the sharing ratio to be reasonable as it left enough credit for paying for tax, reinvesting in the infrastructure, marketing the products, etc. which local people were unlikely to be able to manage themselves.

This is a good example of resilience, in that it shows how their social capital has changed but can remain strong and facilitate collective actions. These communities also gain success in building up relationship with tourism companies, which allows them to perform their role in tourism program and take a reasonable profit sharing.

There were also some other tourism companies in Lak. However, these business units mainly rely on their own assets and have less connection with villagers. The Van Long Company[13] invested in building a simulated traditional house, including

[13] Website of Van Long Company: http://dulichvanlong.com/en

the standard facilities like in a standard tourism accommodation, and much more modern than the communities' actual infrastructure. The Lak Resort[14] is more expensive, as they own a chain of cottages with pretty views of Lak Lake. This private accommodation restricts the participation of the community as homestay suppliers. Both companies even have their own elephants, which are historically just owned by indigenous people. They only hire the local people to perform Gong, drive the elephant, or work as housecleaners. The local people, thus, share a limited amount of the benefits that these companies make from tourism. They are building a boss-hired worker relationship in tourism activities, and many of them perceive *tourism as no more than a simple job, without any excitement and are willing to give up whenever they find another job with a higher salary* (interview at Lak District, 12.2014). Villagers, then, were limited to acting as suppliers as well as employers. In turn, the companies received less support from local people. And their tourism product is far much from the authenticity to attract CBT tourists.

Community Enterprises

Turning to the last case, Lac Duong is the most developed study sites and witnesses the most transformation in their society. Community networks have weakened, and collective activities in these communities are rare and are now completely absent in their tourism activities. That influences both of the bonding relationship among individual households when participating in tourism and the bridging cooperation with outsiders/ tourism companies.

Since local people do not have such a close bonding connection, traditional value consequently is lost, and private enterprises were less likely to be motivated to work with a network and find that it is not necessary to build up their tourism product based on community. Actually, tourism has been developed for a long time now, accompanied by the establishment of a range of tourism entrepreneurs. They include the Lam Dong Joint Stock Tourism Company running a tourism program at Lang Bian Mountain, the Center for Ecotourism and Environmental Education of Bidoup-Nui Ba National Park running an ecotourism program, and the Lam Dong Water Supply Joint Stock Company running a tourism program at Gold Stream. However, the cooperation with private companies does not happen in the same way as this above case. Indigenes express their annoyance when outsiders borrow their culture to conduct tourism services and point out that it hurts their pride.

Ms. Ro Ly, B'NorC Village, Lac Duong District

They build a tourism product based on my culture, but they only briefly learn about actual culture. We do not want to work with them. How can they become manager of these cultural activities while they know so little about the culture? In addition, since they borrow our culture for their business, this should give us the right to good compensation.

---

[14] Website of Lak Resort Company: http://daklaktourist.com.vn/

Meanwhile, local people appeared to be poor in skills and to claim high salary leading companies to refuse to employ them. In a range of tourism companies at Lac Duong District, indigenous people just ran errands or worked in security or sanitation for low pay and *"[were] not interested in their positions, and [did] not pay any attention to developing tourism either*" (interview at Lac Duong District, in 11.2014). The low level of community participation resulted in outsiders reaping more of the economic benefits of tourism.

Generally, the linkage between outside entrepreneurs and villagers was pretty poor. On one side, outsiders developed tourism business based not at all on community. On the other side, local people had full power to make decisions and organize their own business. So far, there are no community tourism enterprises at the study site, but some private companies run by individual villagers grew up in Lac Duong District. Eleven indigenous households with enough assets (human resources, financial capital) have organized Gong groups and run these as independent business. They could also offer supplementary services such as traditional food and beverages *(cơm lam, rượu cần).*

This activity is totally based on common traditional values and is run by local people, so villagers were assumed to have more opportunities to get involved. In fact, the Gong groups were mostly made up of relatives. Apart from food or traditional wine supplied by different households, such activities did not benefit communities much unless they belonged to a Gong group. Moreover, due to the weak community network and lack of cohesion among members, respondents at Lac Duong District revealed a certain level of conflict and competition appearing between Gong groups when developing tourism. The competition is even severer, since local people are so stuck with a very few tourism activities relating to Gong performances that individual households cannot specialize and diversify their tourism participations.

Mrs. Sela, Bon Don Village, Lac Duong District

Our group attracts the most visitors. Therefore, the other groups envy and denigrate us; they fabricate scandals that are even spread in the media. They also cut their prices to draw tourists away from us. The competition is so fierce. They created lots of problems. They convinced the Cultural Department of our district to take over the gong club and promoted it to tourists, then equally divided tourists between the gong groups. By then I had lost my customers, and the tourists were displeased. This model was not effective. In the end, it was dissolved, and each group organized gong performances on their own again.

Not surprisingly, individual interests have recently become more important and carry much more weight than community benefits. Tourism activities were organized and operated individually, rather than as a common product of a community. Local people rejected the concept of equally sharing income (which is derived from equally sharing the number of tourists) and claimed that everything they did was fair in business competition.

When it came to publicity, these small tourism businesses in remote communities usually struggled with the challenges of accessing the market, as they lacked the means to connect directly to tourists. The Gong group owners cared little for developing networks with other on-site companies, whereas they were eager to connect with larger companies located in Dalat and Saigon or more importantly with middlemen who offered them the simple access to the market. So far, middlemen have played an effective role in bridging communities and the tourism market. "*Visitors come here mostly via middlemen*" (interview at Bon Don Village, Lac Duong District, on 14.11.2014). However, the owners were creating unfair competition by paying middlemen commissions to send them tourists. In other words, the appearance of middlemen caused leakages of tourism income and made the competition between groups even fiercer.

Mr. Plyng, Bang Gia Village, Lac Duong District
We were the first gong group here. Following me, other households started their business as gong performances too. This tourism activity blossomed in this area, and then the middlemen appeared. They meet and pick the tourists up from Dalat, or on main roads; they then bring the tourists to Lac Duong, and send them to the groups who give them a generous commission. The quality of the performance does not matter [to middlemen]. Money makes them blind. Now they collect money from tourists first then pay us later. They are the ones who do tourism now, not us anymore.

Middlemen/traders are an essential chain in such tourism industry. However, the relationship between Kinh middlemen and local people in this case is reason for concern, as it is generated based on an unbalanced power. The middlemen take much advantage of the weakness of villagers and earn from their middle position. Local people were aware of the negative sides in the relationship with middlemen, but they lack sufficient assets to reject working with such agents.

The main social networks at three study sites – the connections within communities and the relationship between communities and companies – are described in Fig. 4.5 below. It generally presents a model of three development steps at study sites. The more developed the study site, the more community ties they will lost.

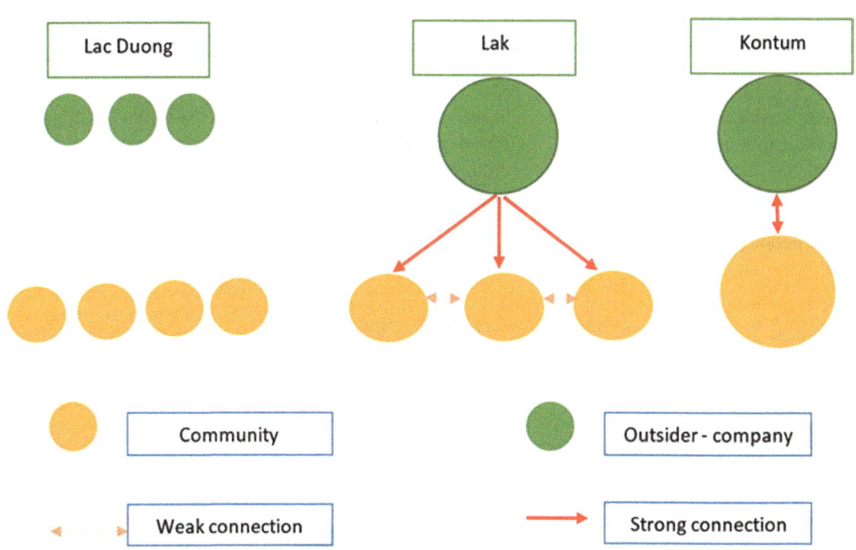

Fig. 4.5 Connection forms in tourism activities at study sites (*Source*: Huỳnh Anh Chi Thái)

Community networks, which used to be a specific characteristic of the indigenous people, have changed due to the eco-social transformation. Simultaneously, different networks between communities and companies have been formed at the study sites, depending on the level of tourism development and the community's ability of self-organization. This kind of bridging network has two sides: supporting the community and making them depend on companies. In some cases, the network with outsiders appears as a social capital compensate which facilitates CBT activities.

4.5 Self-organization and Resilience Livelihood

Since the development of CBT highlights on the community's ownership role, and CBT is examined as a livelihood resilience in this study, "self-organization" is an important attribute that should exist and be promoted. Self-organization is a significant ability in ensuring that the indigenous communities have a sufficient level of autonomy, independence, power, and control, which they can exercise when coping with constraints.

In the specific case of developing CBT, the networks with tourism company reflect the level of indigenous community's self-organization, which interpret into different forms of community involvement/participation in tourism: from passive involvement to active involvement and full participation. Once indigenes are able to participate fully in tourism activities, it is easier for the community to organize their own tourism programs. The scope of community participation is, however, limited depending on the type of tourism enterprises (Ashley and Roe 1998; Thammajinda 2013) (Table 4.6).

Table 4.6 Typology of community participation in tourism

<table>
<tr><th>Passive involvement</th><th>Active involvement</th><th>Full participation</th></tr>
<tr><td rowspan="2">Private businesses organize tourism activities independently from indigenes</td><td>Tourism products are based on indigenous culture</td><td>Enterprise run by indigenous people</td></tr>
<tr><td rowspan="2">Villagers are tourism-service suppliers</td><td>Indigenes design tourism program</td></tr>
<tr><td>Local people join in tourism activities as simple waged labors</td><td>Weakness in access to market, depend on middlemen</td></tr>
<tr><td>Benefit from tourism is very limited</td><td>Sharing benefit based on their supplying</td><td>Conflict in community, benefit leak to outsiders</td></tr>
<tr><td>Private company</td><td>Cooperation with the private sector</td><td>Community enterprise</td></tr>
</table>

Social networks, tourism knowledge and skills, financial capital, marketable

Source: Based on general frame of Ashley and Roe (1998, p.26)

Local people at study sites were participating in tourism at three different levels. The communities in Lac Duong proudly asserted that they could run tourism activities on their own. However, it was not really a community product and did not present a harmonious image of their village, as people preferred to focus on their own benefits and conflicts, which unfortunately inhibited the development of these activities. Moreover, individual businesses did not help local people mobilize their limited resources to compete with rival Kinh-run companies, and their dependence on middlemen led to most of the benefits escaping the community. In contrast, communities in Kontum worked together in peace, kept balance in sharing profit, and set aside a small part for common benefit. However, the tourism program at Kontum was totally organized by outsiders. Meanwhile, tourism activities in Lak District were already shaped by a medium number of companies and visitors in a balanced network of insiders and outsiders. Although their community networks – like at other study sites – were eroded, the new form of kinship was useful enough to keep them out of conflicts and strong enough to demand the respect from tourism companies when participating in tourism and sharing benefits. The model of Lak District is, therefore, proposed as a solution for developing CBT in Central Highlands.

Up to now, the level of self-organization of the local people at study sites is still lower than it needs to be, even in their private businesses. In order to develop CBT, a community network is first required. Bonding social capital makes communities unique from an outside perspective, owing to the value of their culture, norms, traditional customs, and knowledge, and gives community members equilibrium when working together. However, in the first stage of tourism development, the network with companies plays an important role, as it creates chances for a community to participate in the tourism industry and compensate for villagers' asset deficiency. The full participation of indigenous people is reached only when they have sufficient tourism knowledge and financial capital. A sustainable step is defined when the community gains a cooperation built on a balance network of inside and outside communities. That not only strengthens the ability of community organizations but also allows them to take advantages of outside assets.

4.6 Conclusion

4.6.1 Key Facets for a Successful CBT

Although the tourism resources differ among the study sites, they all reveal strength points for these areas to develop tourism products and activities. However, the facets for promoting successful CBT initiatives and creating various CBT products at the study sites are based more on other assets: human and social capital. A lack of quality in certain indicators of human or social capital results in the ineffectiveness of CBT. The success of CBT depends heavily on the community's willingness to get involved. A common weakness of most indigenous communities is their perception of power imbalances. Villagers often perceive themselves as inferior or powerless in comparison to tourism entrepreneurs. As a result, they are hesitant when participating in tourism as active agents and hesitate in taking ownership of CBT. Power imbalances are caused by more than cultural characteristics or social contexts alone. In the process of tourism development, the vulnerability of communities is clearest in their lack of tourism knowledge and skills. The communities with more tourism knowledge are more motivated to implement CBT programs and consider benefit sharing. Moreover, as an alternative livelihood to bring in extra income, CBT should be regarded as a business model more than a community development model. Thus, the entrepreneurial role of the community and their technical skills are required. Also, the key community quality that supports CBT and ensures sustainable tourism practices is community cohesiveness. In particularly, this can help against the constraints they face; collaboration creates a good condition to mobilize assets in a common tourism product. Likewise, solidarity prevents conflict when it comes to sharing benefits, which is still limited at the initial stage of tourism development.

With regard to these above facets, the communities in Lak District have the most potential to run an effective CBT program. Villagers in the Lak District have a number of channels through which they can access tourism knowledge, and some of them already have their own plans to contribute to tourism development. Moreover, the social capital in these villages has kept its structural and cognitive characteristics, although the community networks are now present as a form of kinship network, and the trust and reciprocity among relatives have lessened. Relationships among households also support and are influenced by tourism activities. Individual households at Le and Jun Village, in Lak District, play the role of suppliers for their common tourism products. In addition, the communities in Lak rely on the bridging network with the company, who creates opportunities for villagers to join in with tourism programs and who subsequently become their leaders.

Apart from the modest number of contacts with companies, there are very few connections between communities and external stakeholders. The absence of official organizations and nonprofit organizations clearly indicates that tourism has not yet been developed sufficiently in this area. And indigenous communities are still too vulnerable to develop such specific activities on their own.

4.6.2 Self-organization Is Important for Livelihood Resilience

This study has looked at tourism activities as part of the livelihood typology of indigenes. But most of the decisions related to developing tourism in the Central Highlands at the moment are not initiated by local communities. External stakeholders have recognized the tourism potential in this area and have transformed this potential into tourism products, some specifically into CBT products. However, the community's power is not actually reflected in these activities, and hence communities do not benefit much from them. Depending on different types of tourism enterprises, community participations perform in different levels from active participation at the individual level to passive involvement in which collective income is received or individual salary is paid by the external private business.

Self-organization in CBT is only fully achieved once the actors move from passive to active participation and from individual to collective involvement. At the moment, the communities in the Lac Duong District are the only ones who have actively decided to participate in tourism activities and who could organize tourism activities themselves if they chose to. However, these cases all involve individual entrepreneurs. As such, their activities do not necessary depend on a synergy of collective assets and do not require the full involvement of community members. Instead, the process of tourism development has generated conflict between individual entrepreneurs in this community.

At the initial stage of CBT development, the existence of external agents is inevitable. These might be companies, NGOs, or middlemen. Even in the private entrepreneurs in Lac Duong District, external agents have intervened and influenced the activities as middlemen. External agents do not always create disadvantages since they generate opportunities for communities to participate. The balance of power and the sharing of benefits are aspects that need be improved by the community taking on an actor role.

4.6.3 An Ideal CBT Concept Is Difficult to Transfer into Reality

For vulnerable groups like the indigenous communities in the Central Highlands, collective actions and collective managements are inevitable on their development pathway that enables the actor role of internal communities and prevents them from being too much influenced from outside. However, it has proven difficult to create balance in sharing benefit. So far, the community in Kontum City is the only case where the partners manage to work together in peace, share the tourism benefits, and set aside a small part to village fund. However, the success of tourism in Kontum City was still in its initial stage. There is no guarantee that this business model will maintain its stability when the number of tourists increases or whether higher visitor number will create conflicts and lead to a collapse of the program. The benefit sharing in Lak District is mostly between outsiders and community. Meanwhile, the cooperation within the community in the Lac Duong District has created conflict. Since the most important criteria for a successful CBT program depend on community involvement and a fair distribution of income and benefits to communities, the results of this study clearly show that the process of developing CBT at study sites at that moment needs to be improved.

As a sustainable solution, it is recommended to keep the tourist numbers to a moderate level and to uphold the cooperation in the community and with other stakeholders to continue to compensate for the asset deficiency of the villagers. Moreover, CBT as a resilience livelihood is just a specific case and has not yet become a mass trend (and cannot be practiced by all). Due to the risks that tourism can bring to the local people, for instance, the competition with the other famous destinations and the fluctuation of the tourism market, CBT should be developed on a small-scale and moderate level as an extra income resource. Agriculture is still preferred as the main livelihood for the indigenous people.

References

Ashley C, Roe D (1998) Enhancing community involvement in wildlife tourism: issues and challenges. International Institute for Environment and Development, London, pp 1–41

Bidoup-Nui Ba National Park Center for Ecotourism and Environmental Education (2014) *Community-based ecotourism*. Available at http://bidouptour.com/en/tours.html. Accessed 5 May 2015

Chen H et al (2012) 'Livelihood sustainability and community-based co-management of forest resources in China: changes and improvement', Environ Manage 2011/11/01, 49(1), pp 219–228. doi:https://doi.org/10.1007/s00267-011-9775-4

Claiborne P (2010) Community participation in tourism development and the value of social capital – the case of Bastimentos, Bocas del Toro, Panamá. University of Gothenburg, Gothenburg

Dolezal C (2011) Community-based tourism in Thailand:(Dis-) illusions of authenticity and the necessity for dynamic concepts of culture and power. ASEAS, Aust J South-East Asian Stud Forum South-East Asia 4(1):129–138

García Lucchetti V, Font X (2013) Community based tourism: critical success factors., ICRT occasional paper, vol 27

Harpham T, Grant E, Thomas E (2002) Measuring social capital within health surveys: key issues. Health Policy Plan 17(1):106–111. https://doi.org/10.1093/heapol/17.1.106

Hickey GC (1967) The highland people of South Vietnam: social and economic development. The Rand Corporation, Santa Monica

Highlands Eco Tours (2010) Tours. Available at http://vietnamhighlands.com/aboutus.html. Accessed 10 Aug 2015

Hwang D (2012) Influence of social capital on community-based action in tourism development: a study of social network analysis. University of Illinois at Urbana-Champaign, Champaign

Ifejika Speranza C, Wiesmann U, Rist S (2014) An indicator framework for assessing livelihood resilience in the context of social–ecological dynamics. Glob Environ Chang Elsevier Ltd 28:109–119. doi:https://doi.org/10.1016/j.gloenvcha.2014.06.005

Insititute for Tourism Development Research (2010) Statistics. Available at http://www.itdr.org.vn/en/statistics.html. Accessed 3 Mar 2016

Jones S (2005) Community-based ecotourism: the significance of social capital. Ann Tour Res 32(2):303–324. https://doi.org/10.1016/j.annals.2004.06.007.

Kon Tum Province (2009) Kon Tum qua các thời kỳ lịch sử (Kontum through historical periods). Available at http://www.kontum.gov.vn/pages/lich-su-hinh-thanh.aspx. Accessed 10 June 2016

Lak People's Committee (2014) Statistics. Lak People's Committee, Lak

Lam Dong Province (2009) Qui hoạch khu du lịch Suối Vàng (Planning the Suoi Vang resort). Available at http://www.lamdong.gov.vn/vi-VN/home/quy-hoach-phat-trien/cac-nghanh-linh-vuc/Pages/suoi-vang.aspx. Accessed 30 May 2015

Le, T (2015) Bản Lác, điểm đến thú vị trên bản đồ du lịch Hòa Bình (Lac village, an interesting destination in Hoa Binh Province). Available at http://dulich.vnexpress.net/tin-tuc/tu-van/hanh-trinh/ban-lac-diem-den-thu-vi-tren-ban-do-du-lich-hoa-binh-3133024.html. Accessed 2 June 2016

MacCannell D (1999) The tourist: a new theory of the leisure class. University of California Press, Berkeley

McElwee P (2008) Becoming socialist or becoming Kinh. In: Duncan CR (ed) Civilizing the margins: southeast Asian government policies for the development of minorities. NUS Press, Singapore

Nguyen Ngoc (2008) Một số vấn đề phát triển bền vững ở Tây Nguyên (Some issues regarding to developing the Central Highlands sustainably). Available at http://www.fetp.edu.vn/vn/tin-tuc-su-kien/hoat-dong-tai-truong-fulbright/mot-so-van-de-phat-trien-ben-vung-o-tay-nguyen/. Accessed 8 Aug 2015

Salemink O (2003) The ethnography of Vietnam's Central Highlanders: a historical contextualization, 1850–1990, Anthropology of Asia series. RoutledgeCurzon, London

Suansri P (2003) Community based tourism handbook. Print book. REST, Thailand

Thammajinda R (2013) Community participation and social capital in tourism planning and management in a Thai context. Lincoln University, Lincoln

The Department for International Development (1999) Sustainable livelihoods guidance sheets. Department for International Development, London

Tosun C (2000) Limits to community participation in the tourism development process in developing countries. Tour Manag 21:613–633

UNHCR (2006) Vietnam: situation of indigenous minority groups in the Central Highlands

United States Agency for International Development (2008) Vietnam Central Highlands needs assessment. USAID

Van Breugel L (2013) Community-based tourism: local participation and perceived impacts. A comparative study between two communities in Thailand. Radboud Universiteit Nijmegen, Nijmegen

Vietnam Online (2015) Langbiang Mountain – Dalat. Available at https://www.vietnamonline.com/attraction/langbiang-mountain-dalat.html. Accessed 15 Mar 2016

Vietnam Tours (2016) Ho Chi Minh travel. Available at https://www.tourtovietnam.net/destinations/ho-chi-minh-travel/. Accessed 1 July 2015

Vietnamnet (2015) Danang identified as tourism hub on south-central coast. Available at http://english.vietnamnet.vn/fms/travel/132061/danang-identified-as-tourism-hub-on-south-central-coast.html. Accessed 14 July 2015

Wikipedia (2005a) Đà Lạt. Available at https://en.wikipedia.org/wiki/Da_Lat. Accessed 1 July 2015

Wikipedia (2005b) Nha Trang. Available at https://en.wikipedia.org/wiki/Nha_Trang. Accessed 1 July 2015

Wikipedia (2009) Da Nang. Available at https://en.wikipedia.org/wiki/Da_Nang. Accessed 14 July 2015

World Bank (2011) Vietnam urbanization review: technical assistance report. World Bank, Hanoi. https://doi.org/10.1017/CBO9781107415324.004

Chapter 5
Conclusion and Discussion

Abstract This study aimed to explore the vulnerability context of indigenous people in Vietnam's Central Highlands. The study sheds light on the topic by objectively examining external and internal factors. Land-use change and socioeconomic transformation were investigated as external conditions. Meanwhile, the actor role of indigenous people was examined as an internal factor that enables them to cope with external conditions. The summary of this book highlights the main contributions that this study sought to provide: (1) To what extent do land-use changes and other external conditions influence the livelihood dynamics of indigenous people? (2) What internal factors impede indigenous people's livelihoods, and to what extent do these factors contribute to their ability to cope with constraints?

Keywords Central Highlands in transition · Vulnerability context · Double exposure · Actor role · Sustainable livelihood

5.1 Empirical Findings

Based on the analysis of spatial data and approximately 60 case studies, this study addresses the research questions presented above. The main findings presented in the empirical chapters are synthesized below.

5.1.1 Double Exposure of Vulnerability: The Context of Using Natural Resources

Transformations in the Central Highlands have posed many problems and created a vulnerability context for the indigenous communities. As natural resource users, the communities at the study sites were exposed directly to land constraints, which have had critical and lasting negative impacts on their lives. The livelihood constraint context in this present study was caused by both land scarcity and failures in land

H.A.C. Thái, *Livelihood Pathways of Indigenous People in Vietnam's Central Highlands*, Advances in Asian Human-Environmental Research,
https://doi.org/10.1007/978-3-319-71171-3_5

allocation, as well as how indigenous people have transformed resources (endowment) and derived utility (entitlement) from the land.

A new entitlement map in the Central Highlands and a new power dynamic created an increasing amount of tension for the local people. Initially, a new power relationship emerged from a range of policies which were formulated at the centralized national state level and which ignored all local long-standing informal institutions and customary laws. Another power relationship that hindered the ability of indigenous communities to access natural resources and benefits from them was embedded in the community's relationship with immigrants. A massive flux of migrants who flocked to the Central Highlands in response to the appeal of benefiting from the lucrative coffee market heavily altered the population composition in the area. The population increased from 1,225,000 in 1976 to 4,236,700 in 2000 and reached up to 5,525,800 in 2014. Consequently, the percentage of indigenous people decreased from 69.7% in 1976 to around 26% in 2010. Although migrants and indigenous people had the same rights to access resources, inequality occurred in the different ways people controlled those resources and the distinct processes they used to transform their endowments into entitlements.

Historical background and cultural characteristics partially influenced the process of resource exchange and entitlement mapping. The first pattern influencing highlanders' entitlement mapping was the customary law of land. As they were accustomed to practicing customary law in the past, the new land rights were confusing for the indigenous people, and they were slow to adopt the new land rights. What is more, the failure to recognize high-value croplands exposed indigenous people to risks when selling or transferring their land. Meanwhile, economic motivation and full awareness of land rights resulted in migrants gaining more land claims.

In addition, due to the longer remaining barter system in indigenous communities, the highlanders had less experience with capital flows than lowlanders. This made them hesitant to take risks and invest in high-value crops. Moreover, the persistent existence of self-subsistence systems hindered indigenous people from gaining knowledge of the market and high-tech farming. In contrast, most of the migrants originated from the earlier exploited areas (including the lowlands and northern highlands) and have already acquired skills in growing intensive annual crops and experience in accessing credit to support their new cash crops. Moreover, the lure of high-price coffee crops, along with the compromise of leaving their hometowns, created pull and push factors that motivated migrants to achieve success. Consequently, the migrants' switch from intensive annual crops to cash crops seems to have been less challenging than the shifts made by indigenous people from self-subsistent crops to commercial crops. Ironically, newcomers achieved more success in these highlands than the local people, though they had just arrived without any knowledge about this remote area and many started out with empty hands.

In addition to the power inequality in accessing and using natural resources that put indigenous people at risk of resource constraints, their failure of mobilizing other endowments (e.g., capital, labor) also increased their vulnerability (Leach et al. 1999). Young people were expected to be less influenced by traditional social structures and be more active than their forefathers in utilizing assets besides natural resources. However, there turned out to be even more pressure on them. Due to the system of transferring land to following generations, young people increasingly owned smaller plots of land than previous generations. Education promised to help them cope with land scarcity, enhance indigenous labor quality, and create opportunities to achieve positive livelihood outcomes. Yet, it remained challenging for indigenous students to pursue higher studies and strive for a stable job. In addition to hindrances they faced in pursuing upper-level schooling (long distance to school, excessive school fees, etc.), the low outcome of educational investment discouraged young people from remaining in school. Furthermore, the lack of transparency in job recruitment processes and various corruption issues also thwarted indigenous graduates from successfully applying for a stable job. After going into debt for their children's education, indigenous parents were reluctant to invest even more to buy jobs for their children. Moreover, young indigenous people preferred to mimic the Kinh majority, in particular in the fashion and entertainment trends rather than preserve their specific cultural characteristics. They were seen as a halfway generation, who neither gain enough modern knowledge nor master traditional knowledge. Another pattern influencing indigenous people's livings was their village-centered society. While migration has been a popular choice among rural people seeking better job opportunities in cities, the young highlanders in the Central Highlands were not encouraged by their chances to leave their villages. There was a strong collective spirit in villages, in which indigenes found dignity and comfort. Furthermore, most of them were farmers and untrained laborers; they appeared to be less competitive than others to acquire well-paid jobs in cities. The prospect of economic benefits, therefore, might not compensate the social cost they paid when moving away from their communities.

Not only was it challenging to enrich and effectively utilize the human asset to pursue livelihood strategies, but indigenous people were also likely to fail in their attempts to access affordable credit. Local people preferred to take out loans in the informal sector rather than in the formal system, as the informal process was always available, was much faster, and involved far less (or even no) paperwork. This informal system first appeared in indigenous society along with the growth of the market economy and commercial cash crops and has since penetrated into every indigenous village in a form of grocery stores. They did business based on the shortages of villagers, including the crop inputs (both cash and materials for the production) and market access. Accordingly, indigenous people initially borrowed supplies from stores, then paid for them using their harvested products, and left the task of market access to the stores. In fact, the stores were an essential link in the economic chain in the highland areas, functioning as middlemen. However, these middlemen took much power in deciding prices and interest rate and pushed indigenous people into even more vulnerability. Moreover, the capital dependence trapped many households into a cycle of debt, as their products were often just enough to pay for previous debts, and they would immediately have to take out another loan.

The last noticeable thing regarded to infrastructure, which has been improved over the past few decades, has not benefited the indigenous people's lives much. Many indigenous people failed to mobilize this asset, as they live far removed from the improved infrastructure. Indigenous communities were used to living in isolated forests areas and did not lead commercial lives. As a result, they were pleased to sell their more convenient locations to migrants to get back more profit from land transactions. In return, indigenous farmers had to live and cultivate on a more remote and less fertile land. This restricted them in accessing markets and marginalized them socially. Indigenous people now often have to travel a long way to reach towns and public utilities. Many communities were even inaccessible during the rainy season.

The research findings in this study emphasize that the vulnerability context is caused by the "effectiveness" of people's command over resources. Two main issues have emerged from these findings. First, indigenous people have less power in the contest for resource claims. Second, they are not able to mobilize other assets (e.g., credit, knowledge, skills, infrastructure) in order to make effective use of their land or invest in developing new activities.

5.1.2 Livelihood Pathways as Products of Active Agent Decisions

In general, the livelihood pathways of indigenous people at study sites are not dynamic, as they continue to rely on land and work primarily in agriculture. However, the three study sites have each experienced different levels of transformation in their livelihood pathways. As the southernmost study site, nearby developed zones like the southeast region, (in particular, Ho Chi Minh City), Lac Duong District has been positively influenced. It is the most productive area, as indigenous farmers have already shifted to intensive crops of vegetable and fruit. However, the higher the development stage, the more capital and techniques are required. With regard to this competitive investment, indigenous people were prone to lagging behind, and the development more likely extended the gap between the indigenous people and Kinh majority. The indigenous communities in Kontum City were rather isolated, and many still worked in traditional farming, which did not require much initial investment. In turn, they earned lower outcomes. The last case is Lak District, where the local people have found stable farm work in the paddies. This monoculture needed to be diversified to meet higher daily demands that placed indigenous households in a challenge of land constraints, as almost all arable land had been already occupied by migrants.

On this backdrop of these livelihood pathways, the method of approaching actors was applied at the household level to understand the heterogeneous responses. Moreover, in order to highlight the role of the active agents when coping with pressures, the study used positive deviance methodology to scrutinize successful cases. The analysis of these case studies has yielded several interesting results.

Some of the first positive cases came from the middle-aged or older population groups. The traditional knowledge and experience that they had accumulated supported them well in making a living, regardless of whether they worked in agriculture or in a new sector like tourism. A good knowledge background has helped local people widen their vision and adjust their livelihood pathways. A veteran who had the chance to travel to different places and learn new things outside his village became a successful farmer, thanks to his ability to learn quickly from a Kinh neighbor to invest in high-value crops (pepper, coffee). Another man, who was a teacher under the former regime, became a smart farmer after the reforms and never needed a loan for inputs; he found ways to process his own agricultural products and sell them for a higher output. These deviant cases proved the significance of knowledge. Another deviance was performed by people who work as the state staff, normally teachers or medical staff. This social position brings indigenous people stable incomes, the capital access, and important social networks, which effectively supports their livelihood pathways.

However, the determinants that have led some indigenes to decide on different livelihood pathways are not always explicit. The products of these actors can be understood as a process in which preconditions were already an unconscious and long-standing part. In these cases, actions became the products in response to environmental conditions and social contexts; they after that generated new contexts, providing new platforms for subsequent actions. The success of some households might also be determined by past failures. Livelihood pathways are, therefore, supported by "a chronological series of structuring determination" (Bourdieu 1977, p.86) more than by certain assets.

In general, the determinants of these success cases showed a number of similar assets that play an important role. Given the vulnerability context induced by material asset constraints (in particular, land and capital), this study highlighted more on some immaterial assets (e.g., social and human assets) that brought networks, knowledge, skills, etc. These assets might have compensated for the lack of the others. A good education or applicable experience encouraged indigenes to become engaged in new ventures and put their limited assets to effective use. Meanwhile, social relationships helped to mobilize more assets from outsiders. For indigenous people, informal networks played a significant role, regardless of whether they were derived from a social position, a religious organization, or a traditional community network. While the community network within a village may change to a kinship network and loose some of its strength, we found that the religious networks provided the consistent support, as most indigenous people at the study sites were Christian. In addition, bridging networks with other stakeholders when operationalizing an economic activity was inevitable, particularly with entrepreneurs and middlemen. These findings support the previous finding, as they showed that successful households are the ones who were better able to mobilize other assets into their livelihood pathways.

5.1.3 From Livelihood Resilience to a Sustainable Livelihood

The positive deviance cases above have shown that social and human capital is crucial in order for indigenes to escape vulnerability. These assets were examined again in the context of a specific livelihood to understand the level at which they contribute to resilience. Resilience in this study is regarded as the actors' capacity for adapting to transformation, reliance on their own resources, and their ability to self-organize and reduce risks.

Community-based tourism (CBT) – an off-farm activity that does not depend much on land and requires mobilizing of different assets – was chosen from among a number of different livelihoods to test resilience levels. CBT has recently become a popular trend among small-scale sustainable tourism projects. Initiatives have been successfully conducted in various communities in the Asia-Pacific region. Thailand is one of the most famous CBT destinations, and various CBT options have been developed here since the early 2000s as an alternative form placed at the central interest of national tourism planning. In particular, Thai homestays are good examples for the neighbor country due to its natural beauty, its hospitality, and its commitment to rural development and poverty reduction (Sofield et al. 2004). Research on Thai CBT has emphasized the significant role of local committees in designing successful tourism activities and managing their own economic circumstance in a sustainable manner (Satarat 2010; Pongponrat and Chantradoan 2012). However, Thai CBT has not yet met all expectations, since communities often end up receiving little of the benefits and most of the profits end up going to middlemen. In Thailand, as well as in a number of other developing countries, community participation is still limited, as participatory is a priority to business elites and state elites (Tosun 2000). CBT projects have also been developed in other communities in Malaysia, Lao PDR, the Philippines, Cambodia, and Indonesia. In Vietnam, some first CBT initiatives were launched by SNV.[1] The pilot project was in Sapa, which aimed to establish mechanisms to support the active participation of communities and served as an important lesson for other destinations. However, the excess exposure to mass tourism services in Sapa nowadays may lead to socioeconomic risks (Thai 2015). So far, CBT can be found in some rural areas such as the northern mountainous area, the central coast, and around Mekong Delta. A number of new CBT programs have recently also been established in the Central Highlands.

Undoubtedly, social capital was indispensable for CBT development, which was supposed to be based on a high degree of community control. The stock of bonding social capital formed by village and kinship networks varied among the study sites and contributed to developing CBT at different levels. For instance, while conflicts and competitions have appeared among Gong groups in Lac Duong District, the consistently maintained community networks in Kontum City and Lak District help

[1] SNV: A Netherlands-based international development organization that provides advisory services to governments and local organizations in developing countries (www.snv.org).

maintain the peace among villagers when they work together and support a relative balance in sharing the tourism benefits.

Due to the low-developed stage of tourism at these study sites, the cooperation with outsiders, in particular with entrepreneurs and middlemen, was essential. The bridging social capital was, therefore, necessary to be discussed in this study. The position of indigenes in this bridging network varies depending on the different management styles of the entrepreneurs and influenced the level of community participation in tourism activities, which in turn determines the ratio of sharing the tourism benefits. Although the tourism companies are all located in villages and the attractiveness of their products was more or less based on the indigenous community, very few companies seriously invested in the networks with villagers or enthusiastically offered them jobs and shared benefits with them. Especially in Lac Duong District, there was a range of tourism entrepreneurs who run their businesses without local people's participations. In turn, local people in this area operated their own businesses as Gong performance independently from companies. Middlemen (e.g., tour guides, taxi drivers, photographers) have interfered in their businesses by bringing tourists to Gong groups and seeking commissions. The tourism services in Kontum were less developed. Only one company runs tourism business in indigenous villages at the time we were doing field trip there. At this very early stage of tourism development, reciprocity was a supporting tool between the entrepreneur and community, and there was a relative balance in benefit sharing. The last case was in Lak District. Tourism services in this area were offered by different companies, none of which were owned by local people. Their similar products contributed to shaping a common tourism image of the whole village. However, their different types of management determined the different levels of community participation. The tourism cooperative has maintained the most stable networks with the villagers and given the local people more chances to join in with tourism activities than other private entrepreneurs. This well-organized cooperative played an important role in mobilizing community assets into tourism products.

In general, tourism activities at the study sites were mainly operated and managed by outsiders, which led to the bulk of benefits escaping the community. The participation of local people ranged from partial involvement to full participation. While most communities were willing to join tourism enterprises, they normally refused to adopt a leadership position and therefore have not achieved the active involvement. An explanation for this can be derived from their human capital characteristics. Namely, since the tourism industry calls for specific diverse skills and is labor intensive, most indigenous people have never been trained or have been equipped with only basic tourism knowledge and skills. Language has added even more barriers to their participation. Moreover, local people are not fully aware of the significance of CBT, which respects community values; they, therefore, consider the financial challenges to be greater than their own value assets (village landscape, traditional customs, ethnic characteristics, etc.).

In short, indigenous communities have not yet become self-organized and are not likely to develop tourism services successfully on their own. Tourism, at the moment, has just brought them a small extra income. Tourism activities at the study sites created three different pictures. In contrast to the conflict in the Lac Duong District, cooperation and profit sharing operated well in Kontum City. However, the tourism activities in Kontum were too new at the time of the case studies to make any concrete predictions. It is unclear at present whether or not this "one company, one community" model that operates in Kontum City will remain stable at higher stages of tourism development when the numbers of visitors and tour operators increase. Meanwhile, tourism activities in the Lak District have definitely been shaped by a medium number of companies and visitors. This case should serve as a good example for CBT operations in the Central Highlands. In order to further develop this model sustainably, however, cooperation and the target level of tourism expansion needs to be discussed further.

5.2 Theoretical and Empirical Implications

5.2.1 Theoretical Implications

The existing theory of vulnerability provided a framework to address the research questions in this study. However, these concepts were revisited to further understand the specific vulnerability context. All three components of vulnerability (exposure, sensitivity, and resilience) were deployed to shed light on the situation of an active agent – namely, indigenous people. In other words, the actor concept was integrated into this framework to form the core connection of the three components.

In this study, the term "actor" was understood to have a flexible meaning. It could be conceptualized as single indigenous households or the whole indigenous community sharing similar characteristics. The integration of "actor" into the vulnerability concept contributed two components to this study framework. First, it allowed the view of local people to be shifted from passive victims to active agents. The dynamics of change and the dynamics in the relationship between individual households (agency) and natural-social conditions (structures) were examined. This contributed to overcoming the static nature of the vulnerability concept that emphasizes risk and constraint rather than reaction and adaptation. Furthermore, the term "actor" – in the sense of an indigenous community – was used as the scope of vulnerability, allowing the sphere of internal and external factors to be delineated. External factors therefore not only implied natural conditions, as in previous research, but also included human factors beyond the community and generated a vulnerability context exposed to an internal system.

While investigating the resilience of indigenous communities via a specific livelihood option, the indicators of self-organization and buffer capacity suggested by Ifejika Speranza et al. (2014) were used. Buffer capacity was assessed by means of the available assets. This approach is similar to the sustainable livelihood framework (SLF) (The Department for International Development 1999). However, the SLF falls short on taking local actors' agency into account (Wiesmann et al. 2011).

In addition, in a vulnerability context of land constraints and capital insufficiency, it is unreasonable to stereotypically use the "pentagon assets" of SLF to evaluate all capital without any weight. In particular, the specific livelihood in the present study was community-based tourism (CBT); it was therefore necessary to consider social capital and human capital more than other assets. The symmetric pentagon assets of SLF were distorted with an extent of social and human capital.

Self-organization has been defined in the previous literature as a state where actors determine their own rules (Ifejika Speranza et al. 2014). The findings confirmed that the level of self-organization of a community in the first stage of CBT development was determined by bridging social networks via tourism companies. In this case, self-organization was difficult to assess using material indicators, because, in fact, the communities could not yet run tourism services independently. Therefore, this study expanded the knowledge of self-organization in terms of the position of local actors in relation to entrepreneurs. Different management types in tourism companies were investigated to determine the level of community participation and how its organization contributes to those tourism activities.

5.2.2 Empirical Implications

The empirical implications of this study can provide policymakers and developers with a comprehensive understanding of the dynamics of communities' livelihood pathways. Indigenous people used to be the natural resource users, and they still depend heavily on land today. However, the constraints related to available land do not determine the dynamics of their livelihood and are not the most important capital for local people. Investment into this highlands region should take other assets into account, such as capital and markets, which highlanders have little experience of and inequitable access to. Via cash crops, which are dependent on world markets, local people have been connected to the global economy, although many still live in remote areas. Indigenous farmers mainly sell their products to middlemen; therefore, information about market and market price are important for them when it comes to reducing the inequity in trading with middlemen. Although land constraints do not directly affect the livelihood pathways, the negative consequences of the past land reallocation and land right changes that were made without any consideration of customary law are still felt today. It is necessary to protect indigenous people from unwittingly selling or unfairly exchanging their land. In addition, the strong connection between communities and land (forest land and agricultural land) should be respected to give them a higher responsibility of preserving and a greater likelihood of using these resources.

Overall, agriculture has remained the primary livelihood activity for indigenous people. Thus, any programs for developing the Central Highlands should carefully consider this sector. Theoretically, due to land constraints, local people should diversify into nonagricultural sectors. However, in reality, the successful transformation to agricultural commercial production is a target that is more suitable to their development stage at present. This will serve to accumulate assets for a higher-value livelihood in the future.

Positive deviance among indigenous households has contributed a new picture that is different from the common vulnerability context. Although the successful cases were in the minority, they affirmed the role of active agents and negated the assumption that indigenous people's poverty was intractable. The identification of various positive practices has helped policymakers to formulate suggestions for development plans and policy interventions. These families had positive experiences because they practiced different behaviors from others that resulted from their knowledge assets and social relations. Knowledge is an important asset that fosters indigenous people's ability to engage in high-value livelihoods and ensures its success. Investment should, therefore, be made in education for high-quality output. Meanwhile, social relations compensate for limitations in other assets. Given the failure of farmer cooperatives in these uplands in the past, careful consideration is needed when creating cooperation among indigenous households and between the indigenous community and other stakeholders.

For a specific livelihood like CBT, the recommendations above have again been tested and confirmed. Given the inadequate infrastructure and vulnerability context of indigenous communities, it is suggested that tourism should be developed at a moderate level. In other words, CBT programs should not be promoted as a mass trend to target numbers of visitors. Instead, the knowledge and skills of a community should be in line with the development of CBT first, so that the community can be empowered and fully participate in tourism as owners. CBT should be considered an economic tool that requires local people to enhance their capacity to participate, rather than a developmental tool which allows communities to depend on assets from outsiders. However, a stable development pathway toward vulnerability reduction requires good cooperation that will subsequently bring a relative balance in sharing profits. Cooperation enables small-scale businesses, as well as low-income and low-power indigenes, to mobilize assets. This network should be built within and beyond the community. Future cooperation among CBT participants should aim to meet their mutual needs, and the role of the indigenous community must be regarded as the host of tourism programs.

References

Bourdieu P (1977) In: Gellner E et al (eds) Outline of a theory of practice. Cambrige University Press, Cambridge

Ifejika Speranza C, Wiesmann U, Rist S (2014) An indicator framework for assessing livelihood resilience in the context of social–ecological dynamics. Glob Environ Chang Elsevier Ltd 28:109–119. doi:https://doi.org/10.1016/j.gloenvcha.2014.06.005

Leach M, Mearns R, Scoones I (1999) Environmental entitlements: dynamics and institutions in community-based natural resource management. World Dev 27(2):225–247

Pongponrat K, Chantradoan NJ (2012) Mechanism of social capital in community tourism participatory planning in Samui Island, Thailand. Tourismos 7(1):339–349

Satarat N (2010) Sustainable management of community-based tourism in Thailand. School of Public administration, National Institute of development Administration, Thailand
Sofield BT et al (2004) Sustainable tourism as a tool of eliminating poverty. Australian Cooperative Research Centre for Sustainable Tourism
Thai HAC (2015) Sustainable tourism: community-based tourism in Vietnam's central highlands. In: Kozak N, Kozak M (eds) Tourism development. Cambridge Scholars Publishing, Cambridge, pp 31–44
The Department for International Development (1999) Sustainable livelihoods guidance sheets. Department for International Development, London
Tosun C (2000) Limits to community participation in the tourism development process in developing countries. Tour Manag 21:613–633
Wiesmann U et al (2011) A human actor model as a conceptual orientation in interdisciplinary research for sustainable development. In: Wiesmann U, Hurni H, with an international group of co-editors (eds) Research for sustainable development: foundations, experiences, and perspectives of the Swiss National Centre of competence in research (NCCR) north-south. University of Bern, Bern, pp 231–256

Annex

Fig. A.1 Author, research assistant, and interviewees

H.A.C. Thái, *Livelihood Pathways of Indigenous People in Vietnam's Central Highlands*, Advances in Asian Human-Environmental Research,
https://doi.org/10.1007/978-3-319-71171-3

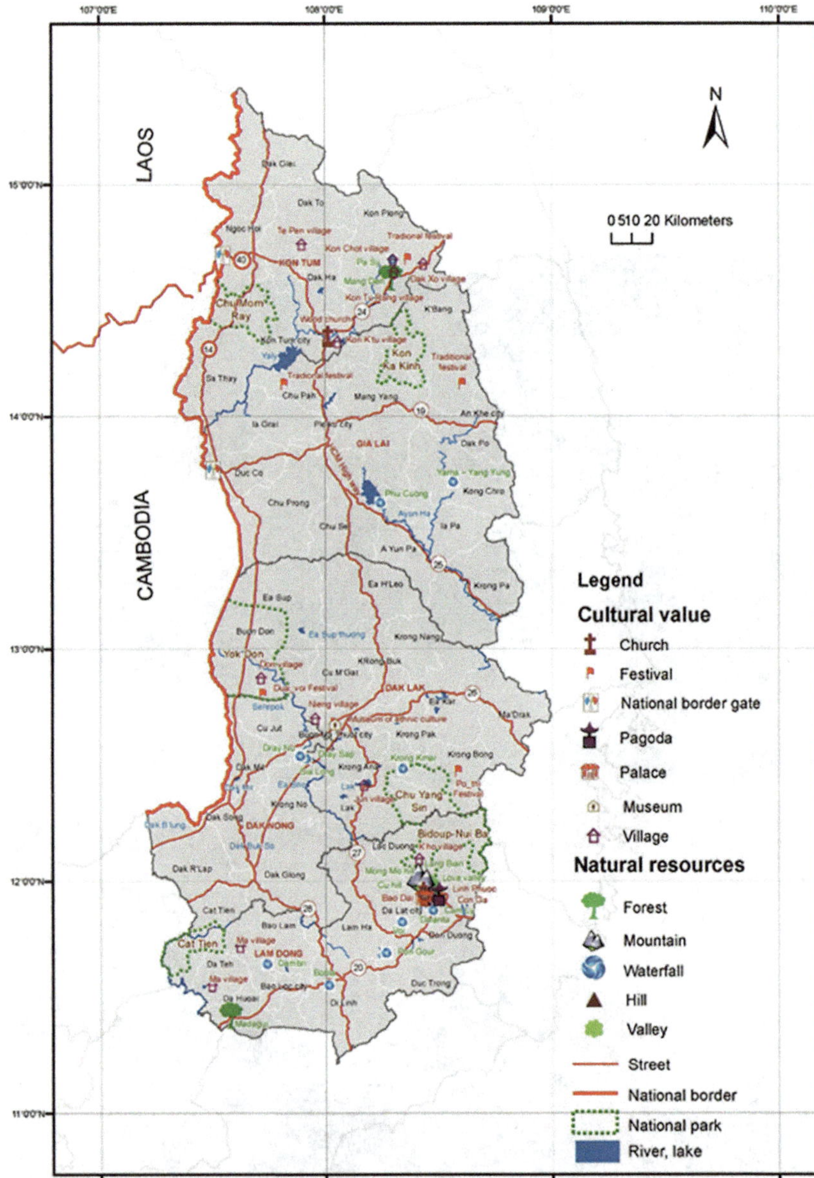

Fig. A.2 Tourism map of Vietnam's Central Highlands (Edited by Huỳnh Anh Chi Thái)

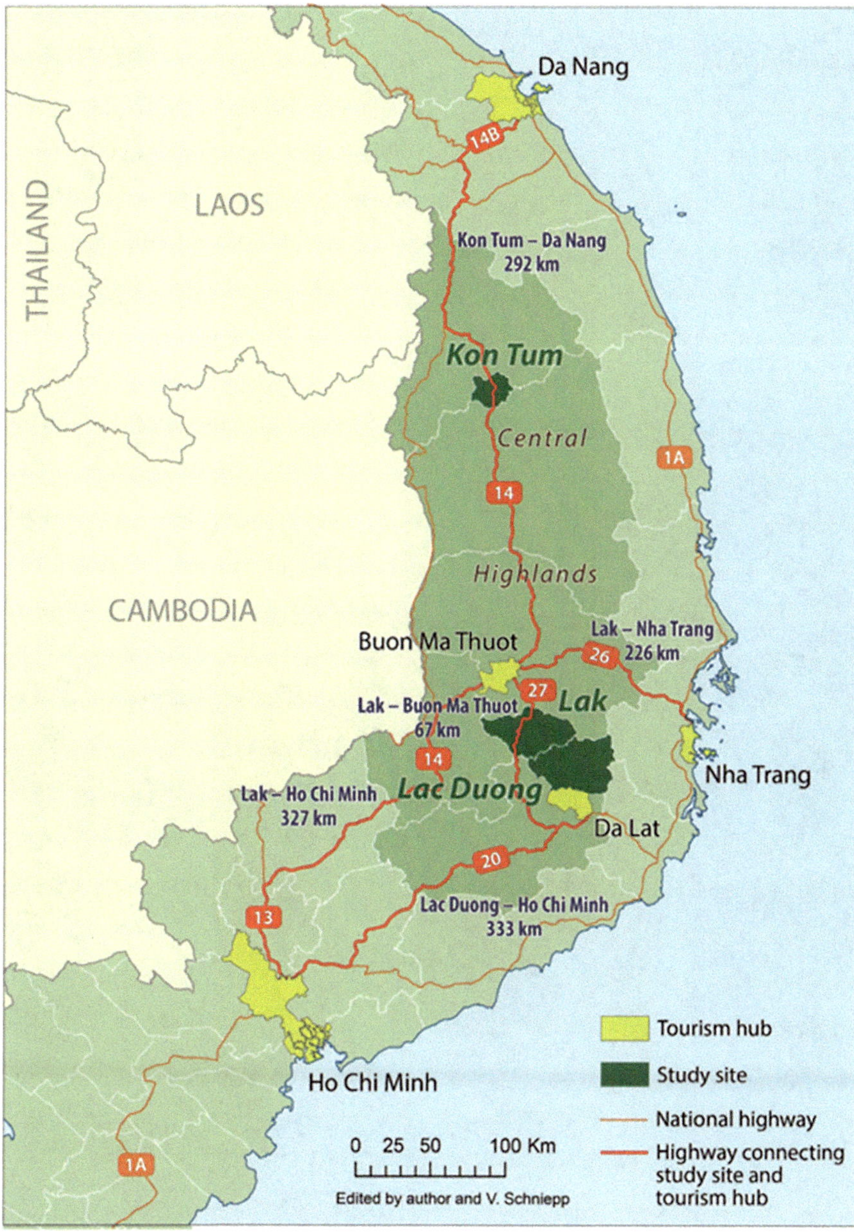

Fig. A.3 Connections of study sites to tourism hub

Index

H.A.C. Thái, *Livelihood Pathways of Indigenous People in Vietnam's Central Highlands*, Advances in Asian Human-Environmental Research,
https://doi.org/10.1007/978-3-319-71171-3